GLENCOE MATH

BUILT TO THE COMMON CORE

AUTHORS
Carter • Cuevas • Day • Malloy
Kersaint • Reynosa • Silbey • Vielhaber

Mc
Graw
Hill
Education

Bothell, WA • Chicago, IL • Columbus, OH • New York, NY

connectED.mcgraw-hill.com

STEM McGraw-Hill is committed to providing instructional materials in Science, Technology, Engineering, and Mathematics (STEM) that give all students a solid foundation, one that prepares them for college and careers in the 21st century.

Send all inquiries to:
McGraw-Hill Education
8787 Orion Place
Columbus, OH 43240

ISBN: 978-0-07-667852-5 (*Volume 1*)
MHID: 0-07-667852-0

Printed in the United States of America.

7 8 9 10 11 QSX 21 20 19 18 17

Common Core State Standards© 2010. National Governors Association Center for Best Practices and Council of Chief State School Officers. All rights reserved.

Understanding by Design® is a registered trademark of the Association for Supervision and Curriculum Development ("ASCD").

CONTENTS IN BRIEF

Everything you need,

anytime, anywhere.

With ConnectED, you have instant access to all of your study materials—anytime, anywhere. From homework materials to study guides—it's all in one place and just a click away. ConnectED even allows you to collaborate with your classmates and use mobile apps to make studying easy.

Resources built for you—available 24/7:

- Your eBook available wherever you are

- Personal Tutors and Self-Check Quizzes to help your learning

- An Online Calendar with all of your due dates

- eFlashcard App to make studying easy

- A message center to stay in touch

Go Mobile!

Visit mheonline.com/apps to get entertainment, instruction, and education on the go with ConnectED Mobile and our other apps available for your device.

v

Chapter 1
Real Numbers

 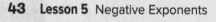

Essential Question

WHY is it helpful to write numbers in different ways?

Real World
p. 89

UNIT PROJECT 103

Music to My Ears

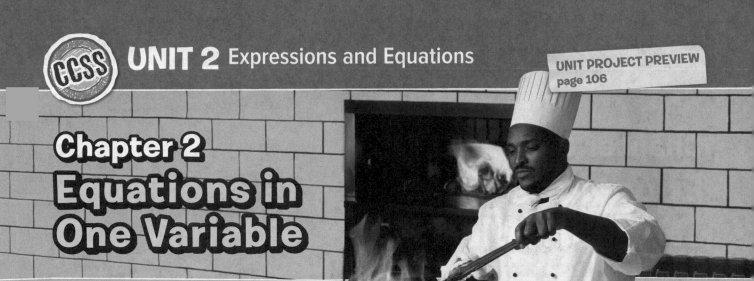

Chapter 2
Equations in One Variable

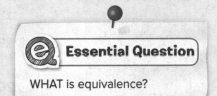

Essential Question

WHAT is equivalence?

Real World
p. 129

Chapter 3
Equations in Two Variables

@ **Essential Question**

WHY are graphs helpful?

Real World

p. 221

Web Design 101

CCSS UNIT 3 Functions

UNIT PROJECT PREVIEW
page 262

Chapter 4
Functions

Essential Question

HOW can we model relationships between quantities?

Real World p. 327

UNIT PROJECT 361

Green Thumb

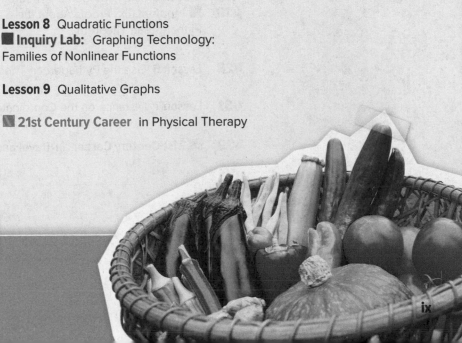

Chapter 5
Triangles and the Pythagorean Theorem

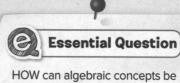

ℯ **Essential Question**

HOW can algebraic concepts be applied to geometry?

Real World
p. 431

x

Chapter 6
Transformations

ⓔ Essential Question

HOW can we best show or
describe the change in
position of a figure?

p. 487

Chapter 7
Congruence and Similarity

Essential Question

HOW can you determine congruence and similarity?

p. 545

(t)Carlos Davila/Photographer's Choice RF/Getty Images, (b)Thinkstock Images/Getty Images   Copyright © McGraw-Hill Education

Chapter 8
Volume and Surface Area

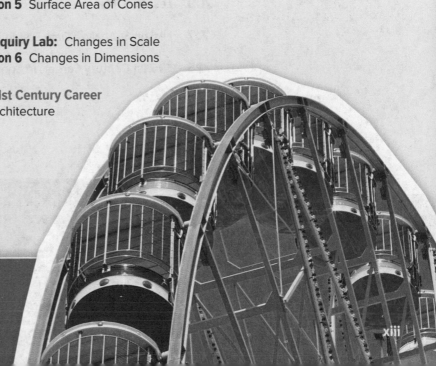

Real World
p. 597

Essential Question

WHY are formulas important in math and science?

UNIT PROJECT 655

Design That Ride!

Chapter 9
Scatter Plots and Data Analysis

p. 701

Essential Question

HOW are patterns used when comparing two quantities?

UNIT PROJECT 731

Olympic Games

(t)©Royalty-Free/Corbis; (c)Karl Weatherly/Photodisc/Getty Images; (b)Thinkstock/Comstock Images/Getty Images

Copyright © McGraw-Hill Education

Common Core State Standards for MATHEMATICS, Grade 8

Glencoe Math, Course 3, focuses on three critical areas: (1) applying equations in one and two variables; (2) understanding the concept of a function and using functions to describe quantitative relationships; (3) applying the Pythagorean Theorem and the concepts of similarity and congruence.

Content Standards

Domain 8.NS

The Number System

- Know that there are numbers that are not rational, and approximate them by rational numbers.

Domain 8.EE

Expressions and Equations

- Work with radicals and integer exponents.
- Understand the connections between proportional relationships, lines, and linear equations.
- Analyze and solve linear equations and pairs of simultaneous linear equations.

Domain 8.F

Functions

- Define, evaluate, and compare functions.
- Use functions to model relationships between quantities.

Domain 8.G

Geometry

- Understand congruence and similarity using physical models, transparencies, or geometry software.
- Understand and apply the Pythagorean Theorem.
- Solve real-world and mathematical problems involving volume of cylinders, cones and spheres.

Domain 8.SP

Statistics and Probability

- Investigate patterns of association in bivariate data.

MP Mathematical Practices

1 Make sense of problems and persevere in solving them.
2 Reason abstractly and quantitatively.
3 Construct viable arguments and critique the reasoning of others.
4 Model with mathematics.
5 Use appropriate tools strategically.
6 Attend to precision.
7 Look for and make use of structure.
8 Look for and express regularity in repeated reasoning.

Track Your Common Core Progress

These pages list the key ideas that you should be able to understand by the end of the year. You will rate how much you know about each one. Don't worry if you have no clue **before** you learn about them. Watch how your knowledge grows as the year progresses!

 I have no clue.   I've heard of it.   I know it!

	Before			After		
8.NS The Number System						
Know that there are numbers that are not rational, and approximate them by rational numbers.						
8.NS.1 Know that numbers that are not rational are called irrational. Understand informally that every number has a decimal expansion; for rational numbers show that the decimal expansion repeats eventually, and convert a decimal expansion which repeats eventually into a rational number.						
8.NS.2 Use rational approximations of irrational numbers to compare the size of irrational numbers, locate them approximately on a number line diagram, and estimate the value of expressions (e.g., π^2).						
8.EE Expressions and Equations						
Work with radicals and integer exponents.						
8.EE.1 Know and apply the properties of integer exponents to generate equivalent numerical expressions.						
8.EE.2 Use square root and cube root symbols to represent solutions to equations of the form $x^2 = p$ and $x^3 = p$, where p is a positive rational number. Evaluate square roots of small perfect squares and cube roots of small perfect cubes. Know that $\sqrt{2}$ is irrational.						
8.EE.3 Use numbers expressed in the form of a single digit times an integer power of 10 to estimate very large or very small quantities, and to express how many times as much one is than the other.						
8.EE.4 Perform operations with numbers expressed in scientific notation, including problems where both decimal and scientific notation are used. Use scientific notation and choose units of appropriate size for measurements of very large or very small quantities (e.g., use millimeters per year for seafloor spreading). Interpret scientific notation that has been generated by technology.						
Understand the connections between proportional relationships, lines, and linear equations.						
8.EE.5 Graph proportional relationships, interpreting the unit rate as the slope of the graph. Compare two different proportional relationships represented in different ways.						
8.EE.6 Use similar triangles to explain why the slope m is the same between any two distinct points on a non-vertical line in the coordinate plane; derive the equation $y = mx$ for a line through the origin and the equation $y = mx + b$ for a line intercepting the vertical axis at b.						

	Before			After		

8.EE Expressions and Equations *continued*

Analyze and solve linear equations and pairs of simultaneous linear equations.

8.EE.7 Solve linear equations in one variable.

a. Give examples of linear equations in one variable with one solution, infinitely many solutions, or no solutions. Show which of these possibilities is the case by successively transforming the given equation into simpler forms, until an equivalent equation of the form $x = a$, $a = a$, or $a = b$ results (where a and b are different numbers).

b. Solve linear equations with rational number coefficients, including equations whose solutions require expanding expressions using the distributive property and collecting like terms.

8.EE.8 Analyze and solve pairs of simultaneous linear equations.

a. Understand that solutions to a system of two linear equations in two variables correspond to points of intersection of their graphs, because points of intersection satisfy both equations simultaneously.

b. Solve systems of two linear equations in two variables algebraically, and estimate solutions by graphing the equations. Solve simple cases by inspection.

c. Solve real-world and mathematical problems leading to two linear equations in two variables.

8.F Functions

Define, evaluate, and compare functions.

8.F.1 Understand that a function is a rule that assigns to each input exactly one output. The graph of a function is the set of ordered pairs consisting of an input and the corresponding output.

8.F.2 Compare properties of two functions each represented in a different way (algebraically, graphically, numerically in tables, or by verbal descriptions).

8.F.3 Interpret the equation $y = mx + b$ as defining a linear function, whose graph is a straight line; give examples of functions that are not linear.

Use functions to model relationships between quantities.

8.F.4 Construct a function to model a linear relationship between two quantities. Determine the rate of change and initial value of the function from a description of a relationship or from two (x, y) values, including reading these from a table or from a graph. Interpret the rate of change and initial value of a linear function in terms of the situation it models, and in terms of its graph or a table of values.

8.F.5 Describe qualitatively the functional relationship between two quantities by analyzing a graph (e.g., where the function is increasing or decreasing, linear or nonlinear). Sketch a graph that exhibits the qualitative features of a function that has been described verbally.

8.G Geometry

Understand congruence and similarity using physical models, transparencies, or geometry software.

8.G.1 Verify experimentally the properties of rotations, reflections, and translations:

a. Lines are taken to lines, and line segments to line segments of the same length.

b. Angles are taken to angles of the same measure.

c. Parallel lines are taken to parallel lines.

CCSS For more about the Common Core State Standards go to **commoncoresolutions.com**.

xvii

	Before			After		

8.G Geometry *continued*

8.G.2 Understand that a two-dimensional figure is congruent to another if the second can be obtained from the first by a sequence of rotations, reflections, and translations; given two congruent figures, describe a sequence that exhibits the congruence between them.

8.G.3 Describe the effect of dilations, translations, rotations, and reflections on two-dimensional figures using coordinates.

8.G.4 Understand that a two-dimensional figure is similar to another if the second can be obtained from the first by a sequence of rotations, reflections, translations, and dilations; given two similar two-dimensional figures, describe a sequence that exhibits the similarity between them.

8.G.5 Use informal arguments to establish facts about the angle sum and exterior angle of triangles, about the angles created when parallel lines are cut by a transversal, and the angle-angle criterion for similarity of triangles.

Understand and apply the Pythagorean Theorem.

8.G.6 Explain a proof of the Pythagorean Theorem and its converse.

8.G.7 Apply the Pythagorean Theorem to determine unknown side lengths in right triangles in real-world and mathematical problems in two and three dimensions.

8.G.8 Apply the Pythagorean Theorem to find the distance between two points in a coordinate system.

Solve real-world and mathematical problems involving volume of cylinders, cones, and spheres.

8.G.9 Know the formulas for the volumes of cones, cylinders, and spheres and use them to solve real-world and mathematical problems.

8.SP Statistics and Probability

Investigate patterns of association in bivariate data.

8.SP.1 Construct and interpret scatter plots for bivariate measurement data to investigate patterns of association between two quantities. Describe patterns such as clustering, outliers, positive or negative association, linear association, and nonlinear association.

8.SP.2 Know that straight lines are widely used to model relationships between two quantitative variables. For scatter plots that suggest a linear association, informally fit a straight line, and informally assess the model fit by judging the closeness of the data points to the line.

8.SP.3 Use the equation of a linear model to solve problems in the context of bivariate measurement data, interpreting the slope and intercept.

8.SP.4 Understand that patterns of association can also be seen in bivariate categorical data by displaying frequencies and relative frequencies in a two-way table. Construct and interpret a two-way table summarizing data on two categorical variables collected from the same subjects. Use relative frequencies calculated for rows or columns to describe possible association between the two variables.

Mathematical Practices Handbook

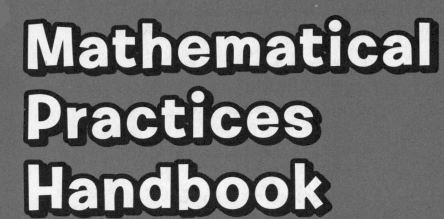

 Essential Question

WHAT practices help me develop and demonstrate mathematical understanding?

 Mathematical Practices

The standards for mathematical practice will help you become a successful problem solver and to use math effectively in your daily life.

What You'll Learn

MP Throughout this handbook, you will learn about each of these mathematical practices and how they are integrated in the chapters and lessons of this book.

① **Focus on Mathematical Practice**
Persevere with Problems

② **Focus on Mathematical Practice**
Reason Abstractly and Quantitatively

③ **Focus on Mathematical Practice**
Construct an Argument

④ **Focus on Mathematical Practice**
Model with Mathematics

⑤ **Focus on Mathematical Practice**
Use Math Tools

⑥ **Focus on Mathematical Practice**
Attend to Precision

⑦ **Focus on Mathematical Practice**
Make Use of Structure

⑧ **Focus on Mathematical Practice**
Use Repeated Reasoning

Place a checkmark below the face that expresses how much you know about each Mathematical Practice. Then explain in your own words what it means to you.

😞 I have no clue.   🙁 I've heard of it.   🙂 I know it!

Mathematical Practices

Mathematical Practice	😞	🙁	🙂	What it Means to Me
①				
②				
③				
④				
⑤				
⑥				
⑦				
⑧				

Persevere with Problems

How do I make sense of a problem?

Making and using a step-by-step plan to solve a problem is like using directions to build a piece of furniture. If you follow the directions correctly, there is a good chance you will end up with a solid piece of furniture. Once you understand the meaning of the problem, you can decide what strategy will work best to solve it. You might try several strategies and then ask yourself, "Does this make sense?"

You have already used the four-step problem-solving plan in previous courses. Complete the graphic organizer that shows the four steps to solve the given problem.

Mathematical Practice 1

Make sense of problems and persevere in solving them.

Of the 480 students at Lincoln Middle School, one third have traveled overseas. Of these, 15% have been to Australia. How many students have not been to Australia?

Step 1.   Understand

What are the facts?

_____

_____

_____

Step 2.   Plan

What strategy will you use to solve the problem above?

_____

_____

_____

Step 3.   Solve

Solve the problem. Show your steps below.

_____

_____

_____

Step 4.   Check

How do you know your answer is reasonable?

_____

_____

_____

It's Your Turn!

Solve each problem by using the four-step problem-solving model.

1. About fifty percent of the population of Alaska lives within a 50-mile radius of Anchorage. If the total area of Alaska is 586,412 square miles, about what percent of the total land area is within 50 miles of Anchorage?

    Understand What are you asked to find? Is there any information you will not use?

    _____

    _____

    Plan How will you solve this problem?

    _____

    _____

    Solve Solve the problem. Show your steps below. What is the solution?

    _____

    _____

    Check Does your answer make sense?

    _____

Check

Solve the problem using a different strategy to check your work.

_____

_____

_____

_____

2. The first three molecules for a certain family of hydrocarbons are shown. How many hydrogen atoms (H) are in a molecule containing 6 carbon atoms (C)?

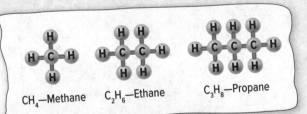

CH_4—Methane C_2H_6—Ethane C_3H_8—Propane

_____

Find it in Your Book!

MP **Persevere with Problems**

Look at Chapter 1. Give an example of where Mathematical Practice 1 is used. Then explain why your example represents this practice.

_____

_____

_____

_____

Reason Abstractly and Quantitatively

What does it mean to reason abstractly and quantitatively?

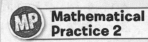

MP **Mathematical Practice 2**

Reason abstractly and quantitatively.

In math, we solve real-world problems where numbers and variables in an equation represent concrete objects. This involves thinking quantitatively.

Suppose you are given a $25 gift card to an online music store. Each song costs $1.95 to purchase and download. How many songs can you buy?

1. What values in the problem do we already know?

_____

2. What are we trying to find?

_____

3. What symbol can we use to represent the unknown value?

_____

Now that the problem is broken down into known and unknown values, we can manipulate the symbols in order to solve the problem. This is thinking abstractly.

4. Write an equation to solve the problem. Explain what each quantity or symbol represents.

_____

_____

5. Use your equation to solve the problem and label your solution. Explain the meaning of the solution.

_____

_____

_____

Write and solve an equation for each of the following.

6. You are in the pit crew for a driver at a Nascar race. The gas weighs 5.92 pounds per gallon. Your driver uses 0.25 gallon per lap. With 42 laps to go, you put 60 pounds of fuel in the tank of the car. Will your driver finish the race at the same rate without more gas?

   a. What values do we already know? What are we trying to find?

   _____

   _____

   b. Write an equation to find the number of gallons in 60 pounds of fuel.

   _____

   c. Use the equation to solve the problem and explain the meaning of the solution.

   _____

   _____

Number of Students, s	Total Cost, c ($)

7. A class trip is scheduled for an amusement park. Group admission prices are $31 per student. Parking is $18 per bus.

   a. Complete the table to show the total cost of 10, 20, 30, and 40 students and two buses.

   b. Write an equation to show the total cost c if two buses

   transport s students to the park. _____

   c. There are a total of 78 students attending on two buses. What is the total cost? Label your solution and explain its meaning.

   _____

   _____

Find it in Your Book!

MP **Reason Abstractly**

Look at Chapter 2. Give an example of where Mathematical Practice 2 is used. Then explain why your example represents this practice.

_____

_____

_____

_____

Construct an Argument

How do I construct a viable argument in math class?

MP **Mathematical Practice 3**

Construct viable arguments and critique the reasoning of others.

Suppose your friend told you that his rectangular flatscreen T.V. has congruent diagonals, simply because it was rectangular. How could you ask your friend to justify his argument? You could use inductive reasoning or deductive reasoning. *Inductive reasoning* uses examples to draw conclusions, while *deductive reasoning* uses definitions, rules, or facts.

1. How could you use *inductive reasoning* to justify why the following statement is true?

   All rectangles have diagonals that are congruent.

   _____

   _____

   _____

2. How could you use *deductive reasoning* to justify why the following statement is false?

   Each angle of an equilateral triangle measures 90°.

   _____

   _____

   _____

   _____

3. Complete the graphic organizer to show that the statement below is *sometimes* true.

The product of two integers is positive.

Example when it is true:

Example when it is false:

It's Your Turn!

For each of the following statements, determine if the statement is *always*, *sometimes*, or *never* true. Justify your response using examples or counterexamples.

4. The sum of two rational numbers is a rational number.

_____

5. The sum of two odd numbers is an odd number.

_____

_____

6. The volume of a pyramid is less than the volume of a prism with the same size base.

_____

_____

7. The United States flag is in the shape of a rectangle with a width to length ratio of about one to two. The length of the blue rectangle *b* is about three-fourths of the entire width of the flag. Suppose the length of the blue rectangle is 48 inches. Kiely thinks the length of the entire flag should be about 72 inches. Is she correct? Justify your response.

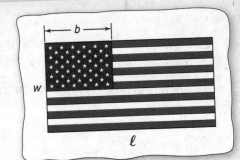

_____

_____

ⓕind it in Your Book!

MP **Construct an Argument**

Look at Chapter 1. Give an example of where Mathematical Practice 3 is used. Then explain why your example represents this practice.

_____

_____

_____

Model with Mathematics

How does math fit into your future?

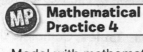

No matter what career path you choose, you are sure to use math in your job or career. Graphic organizers arrange ideas so that you can make informed decisions. Using and understanding models such as graphs, tables, and diagrams helps you to simplify a complicated situation and to identify important quantities in a real-life situation.

Suppose you are a doctor or a nurse. A prescription directs a patient to take 2.5 cc (cubic centimeters) of a medicine per 50 pounds of body weight.

1. What skill(s) would you use to see how much medicine you should give to a 125 pound person?

_____

2. How much medicine would the 125 pound patient need?

_____

3. What career path interests you? Research that career and complete the graphic organizer below.

Education Required

_____
_____
_____

Career: _____

_____
_____
_____

How is math used in this career?

Use the given tools to solve each problem.

Week, w	Total Saved, s ($)

4. You are saving money to buy a new game system. You received $50 as a birthday gift from your grandparents. You want to save $25 a week from mowing lawns.

   a. **Tables** Complete the table to show the total amount saved after 1, 2, 3, 4, and 5 weeks.

   b. **Symbols** Write an equation to show the total amount saved s after w weeks. _____

   c. **Algebra** Use the equation to determine the total amount saved after 17 weeks. _____

Use the table for Exercises 5 and 6.

5. Mrs. Gutierrez hired a party planner to plan Carina's quinceañera. There will be 125 guests and she wants to offer appetizers and a buffet dinner. What is the cost, before tax, for the party?

   _____

Polly's Perfect Parties			
Cost of Food (per person)		Cost of Extras	
Appetizers	$9.20	Hall	$250
Buffet	$18.30	Linens	$15 per table
Sit-down Dinner	$25.75	Table and Chair Rental (seats 8)	$60 per table

6. There is a $7\frac{1}{2}$% sales tax added to the party bill. Mrs. Gutierrez also wants to add an 18% tip for the servers. This will be figured before tax is added. What will be the total cost of the party?

   _____

Find it in Your Book!

MP **Model with Mathematics**

Look at Chapter 1. Give an example of where Mathematical Practice 4 is used. Then explain why your example represents this practice.

_____

_____

_____

_____

Use Math Tools

How do I use tools and strategies in math class?

MP Mathematical Practice 5

Use appropriate tools strategically.

Sometimes using math tools and strategies helps make solving problems easier if you know which tool to use in a given situation. Math tools are physical objects you use when solving problems. Paper and pencil, technology, or calculators are examples of tools.

1. List three other tools you can use to solve math problems.

_____

_____

Math strategies are more like skills or the ability to apply your math knowledge. Some math strategies are mental math, number sense, estimation, drawing a diagram, or solving a simpler problem.

2. List three other strategies you can use to solve math problems.

_____

_____

3. Complete the graphic organizer.

Problem	Tool	Strategy
You want to leave a 20% tip for your server.		
You want to determine how long it will take to drive from Austin to Dallas.		
You are stuck while in the middle of solving an equation.		

It's Your Turn!

List the tools or strategies you would use to solve each problem.
Then solve the problem.

4. A pre-election survey was taken in Ms. Bowen's homeroom.
The results for class president are shown in the table.

Class President	
Elan	10
Karam	8
Magdalena	20
Sara	12

   a. Based on the survey, if there are 850 students in the
   8th grade, how many votes will Magdalena get?

   _____

   b. A candidate needs to receive at least 51% of the votes to
   win the election. If every student votes, how many more
   votes would Magdalena need to win?

   _____

5. A walking path around a lake is in the shape
of a pentagon like the one shown. If Natalie
wants to walk $4\frac{1}{2}$ miles, how many times
does she need to walk around the lake?

   _____

   _____

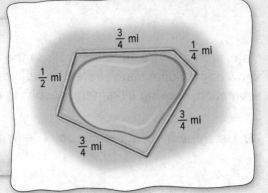

6. Write a word problem that requires the use of a protractor, a calculator, and
one strategy, like mental math or estimation. Find the solution to your
problem and explain how you used the tools to solve it.

   _____

   _____

Find it in Your Book!

MP Use Math Tools

Look at Chapter 1. Give an example
of where Mathematical Practice 5
is used. Then explain why your
example represents this practice.

Attend to Precision

What does it mean to be precise?

Communication is important to our daily life, whether it's in school, sports, at home, or hanging out with friends. If you can't clearly express your thoughts, no one will understand what you mean! Math also requires clear and precise communication by using labels, appropriate symbols, and clear definitions.

Suppose you and your brother want to paint two walls in your bedroom a new color. Your bedroom is 12 feet 5 inches long, 14 feet 8 inches wide, and has an 8-foot ceiling height.

1. What skill(s) would you use to see how much paint you need?

   _____

2. What information do you need to know in order to make your calculations?

   _____

   _____

You are painting two walls that are perpendicular to each other. They do not have doors or windows on them. A gallon of paint covers about 350 square feet.

3. What is the area of wall space you will be painting? Label your answer.

   _____

4. How precise does the area need to be to determine how much paint you will need? Round the area and explain why you rounded to the place value you chose.

   _____

   _____

   _____

5. How many gallons of paint do you need? Round to an appropriate place value and label your answer. Explain your rounding.

   _____

   _____

6. Turn to page 7 in your text. Find the vocabulary term *rational number* and complete the graphic organizer for that term.

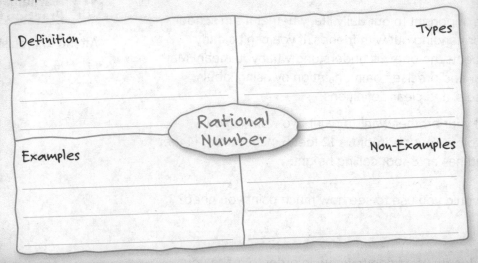

Definition

Types

Rational Number

Examples

Non-Examples

7. Model trains come in different scales. The ratio for an HO scale train is 1:87, while the ratio for a Z scale train is 1:220. Suppose a Z scale model of a steam engine is 62 millimeters long. What is the length of the HO scale model of the same engine? To what place value should you round? Explain your reasoning.

Find it in Your Book!

MP **Attend to Precision**

Look at Chapter 1. Give an example of where Mathematical Practice 6 is used. Then explain why your example represents this practice.

What does it mean to use structure in math?

MP Mathematical Practice 7

Look for and make use of structure.

When you use structure in math, you might apply properties to solve equations or you might examine patterns in tables and graphs to describe relationships.

1. The table shows the diameters of several flying discs. Use the relationship between the radius and diameter of a circle to complete the table. Round to the nearest tenth.

Diameter (cm)	Radius (cm)	Circumference (cm)	Area (cm²)
20			
22			
25			

2. Describe the relationship between the diameter and radius of a circle. _____

3. Describe the relationship between the circumference and diameter of a circle. _____

_____

4. Complete the graphic organizer by writing a formula in each box that shows the relationship between each term.

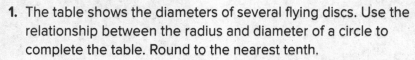

Diameter and Radius

Radius and Circumference

Diameter and Area

Suppose you are training for a marathon. A marathon is 26.2 miles long. You can run 3 miles in 16 minutes.

5. At this rate, how many miles can you run in one hour?

_____

6. Complete the table and plot the points to make a line graph.

Time (h)	Distance (mi)
1	
2	
3	

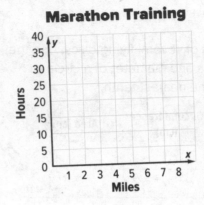

Marathon Training

7. Write an equation that shows the relationship between distance and time.

_____

8. Estimate how long it will take to complete the marathon.

_____

Find it in Your Book!

MP Make Use of Structure

Look at Chapter 1. Give an example of where Mathematical Practice 7 is used. Then explain why your example represents this practice.

Use Repeated Reasoning

What does it mean to look for repeated reasoning?

Problems can often be solved by finding patterns or repeated processes. Sometimes you can even create shortcuts to solve a problem once you understand the pattern. For example, multiplication is a shortcut for repeating the same addition over and over.

Suppose you have a garden with a length of 6 feet and a width of 4 feet and you want to increase its size. Before making any changes, do some math!

1. What is the perimeter of the garden? _____

   the area? _____

2. If you double the dimensions of the garden, what is the new

   perimeter? _____ new area? _____

3. What number can you multiply the original perimeter by to

   find the new perimeter? _____ What number can you

   multiply the original area by to find the new area? _____

Oh no, the increased size of the garden is too big! Using the original dimensions of the garden, you increase the length to 9 feet and the width to 6 feet.

4. What is the new perimeter? _____ new area? _____

5. What number can you multiply the original perimeter by to find the

   new perimeter? _____ What number can you multiply

   the original area by to find the new area? _____

6. Try other changes in the dimensions of the garden to find the new
   perimeter and area of the garden.

   _____

   _____

7. James is mixing orange juice and apple juice in a ratio of 3 to 4 to make a fruit punch. He wants to make 35 cups of the punch. To determine how many cups of each juice he needs, he started making a table. Complete the table to find how many cups of each juice he will need. Then explain a shortcut you could use to solve the problem.

Orange Juice	Apple Juice	Total Cups
3	4	7
6	8	14
9	12	21

_____

_____

_____

8. Savannah's parents are going to pay her for doing chores 6 days a week and they offer her two payment plans.

Option A						
Day 1	Day 2	Day 3	Day 4	Day 5	Day 6	Total
$3	$6	$9				

Option B						
Day 1	Day 2	Day 3	Day 4	Day 5	Day 6	Total
$0.75	$1.50	$3.00				

Complete the table to determine which is the better option for Savannah to choose. Explain the pattern for each option.

_____

_____

_____

_____

Find it in Your Book!

MP **Use Repeated Reasoning**

Look at Chapter 1. Give an example of where Mathematical Practice 8 is used. Then explain why your example represents this practice.

_____

_____

_____

Use the Mathematical Practices

Solve.

You are boxing and wrapping gifts for a club fundraiser. The charge to wrap a gift in the shape of a rectangular prism is shown in the table.

Total Surface Area	Cost
up to 35 in²	$5
36–54 in²	$8
over 55 in²	$12

a. Marina wrapped three different boxes with measurements shown in the table. Complete the table with the cost per box and the cost per square inch. Which box has the least cost per square inch? _____

Box	height in.	width in.	length in.	Cost to Wrap	Cost per Square Inch
A	2	4	3		
B	2	5	6		
C	2	3	2		

b. Which of those boxes has the least cost per cubic inch? Explain.

_____

_____

_____

_____

Determine which mathematical practices you used to determine the solution. Shade the circles that apply.

Which **MP Mathematical Practices** did you use?
Shade the circle(s) that applies.

① Persevere with Problems   ⑤ Use Math Tools

② Reason Abstractly   ⑥ Attend to Precision

③ Construct an Argument   ⑦ Make Use of Structure

④ Model with Mathematics   ⑧ Use Repeated Reasoning

Reflect

 Answering the Essential Question

Use what you learned about the mathematical practices to complete the
graphic organizer. List three practices that help you best demonstrate
mathematical understanding. Then give an example for each practice.

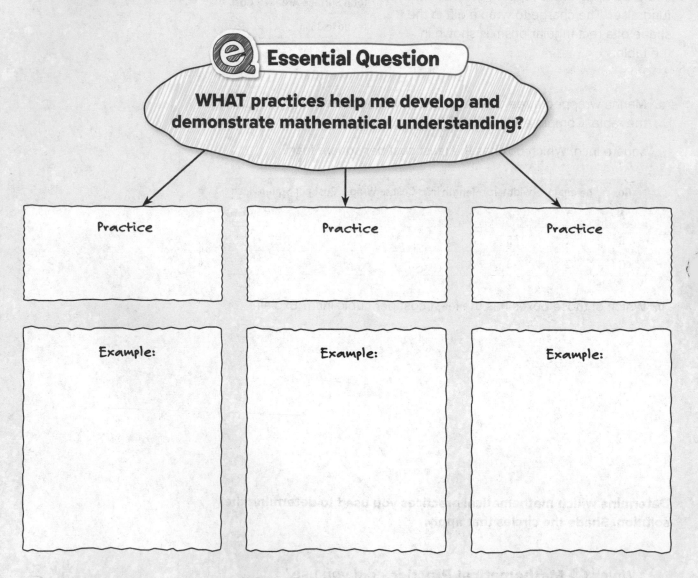

Essential Question

WHAT practices help me develop and
demonstrate mathematical understanding?

Practice

Practice

Practice

Example:

Example:

Example:

Answer the Essential Question. WHAT practices help me develop and
demonstrate mathematical understanding?

_____

_____

_____

UNIT 1

The Number System

Essential Question

HOW can mathematical ideas be represented?

Chapter 1
Real Numbers

Rational numbers can be used to approximate the value of irrational numbers. In this chapter, you will perform operations on monomials and numbers written in scientific notation. You will then use rational approximations to estimate roots and to compare real numbers.

 Music to My Ears Listening to music can be both fun and relaxing. But did you know that many interesting relationships exist between math and music? Even the ancient Greek mathematician Pythagoras observed and wrote about many of these relationships.

At the end of Chapter 1, you'll complete a project to find how math and music are connected. But for now, write about the connections between math and music in the space provided.

 Math and Music

Chapter 1

Real Numbers

 Essential Question

WHY is it helpful to write numbers in different ways?

 Common Core State Standards

Content Standards
8.NS.1, 8.NS.2, 8.EE.1, 8.EE.2, 8.EE.3, 8.EE.4

 Mathematical Practices
1. 3, 4, 5, 6, 7, 8

Math in the Real World

Space The average distance from Earth to the Moon is about 384,403 kilometers. The Sun is the closest star to Earth and is about 150 million kilometers away. The next closest star is Proxima Centauri which is about 4.22 light years away from Earth.

A light year is defined as 9,461 billion kilometers. Find and label the distance in kilometers from Earth to Proxima Centauri.

 **Study Organizer**

 Cut out the Foldable on page FL3 of this book.

2 Place your Foldable on page 100.

 Use the Foldable throughout this chapter to help you learn about real numbers.

3

Vocabulary

base	perfect cube	repeating decimal
cube root	perfect square	scientific notation
exponent	power	square root
irrational number	radical sign	terminating decimal
monomial	rational number	

Use a Mnemonic Device

When a mathematical expression has a combination of operations, the order of operations tells you which operation to perform first. How can you remember the orders easily? A mnemonic device is a verse or phrase to help you remember something.

In this case, it is *Please Excuse My Dear Aunt Sally*. On each rung of the ladder, fill in the operation that the mnemonic device represents. Then evaluate the numerical expression step-by-step.

$$3(5 - 15)^2 - 7 \cdot 3 + 24 \div 6$$

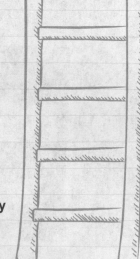

Please _____

Excuse _____

My Dear _____

Aunt Sally _____

Place a checkmark below the face that expresses how much you know about each concept. Then scan the chapter to find a definition or example of it.

 I have no clue. 😕 I've heard of it. 🙂 I know it!

Integers				
Concept	🙁	😕	🙂	Definition or Example
irrational numbers				
multiplying and dividing monomials				
negative exponents				
rational numbers				
scientific notation				
square roots				

When Will You Use This?

Here are a few examples of how real numbers are used in the real world.

Activity 1 Use the Internet or another media source to find a description of the metric system. How do you convert one measurement to another using the metric system?

_____

_____

Activity 2 Go online at **connectED.mcgraw-hill.com** to read the graphic novel **Measuring Up**. One nanometer is equal to how many meter(s)?

_____

Jacob and Sarah in

Measuring Up

Looks like you're 1.52 meters tall.

Are You Ready?

Try the Quick Check below.
Or, take the Online Readiness Quiz.

Example 1

Find 5 · 4 · 5 · 4 · 5.

$$5 \cdot 4 \cdot 5 \cdot 4 \cdot 5 = 4 \cdot 4 \cdot 5 \cdot 5 \cdot 5$$
$$= (4 \cdot 4) \cdot (5 \cdot 5 \cdot 5)$$
$$= 16 \cdot 125$$
$$= 2{,}000$$

Example 2

Find the prime factorization of 60.

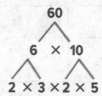

The prime factorization of 60 is
$2 \times 2 \times 3 \times 5$.

Quick Check

Simplify Expressions Find each product.

1. $2 \cdot 2 \cdot 4 \cdot 4 \cdot 4 =$ _____

2. $(-8)(-8)(5)(5)(-8) =$ _____

3. The students at Hampton Middle School raised $8 \cdot 8 \cdot 2 \cdot 8 \cdot 2$ dollars to help build a new community center. How much money did they raise?

_____

Prime Factorization Find the prime factorization of each number.

4. 36 _____

5. 24 _____

6. 18 _____

7. 100 _____

8. 121 _____

9. −42 _____

Which problems did you answer correctly in the Quick Check? Shade those exercise numbers below.

① ② ③ ④ ⑤ ⑥ ⑦ ⑧ ⑨

Rational Numbers

Vocabulary Start-Up

 Vocab abc

Numbers that can be written as a comparison of two integers, expressed as a fraction, are called **rational numbers**.

Complete the graphic organizer.

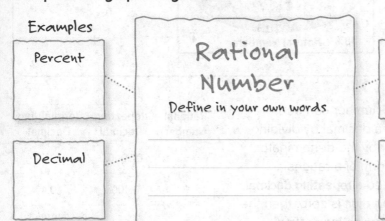

Examples

Percent

Decimal

Rational Number
Define in your own words

Examples

Fraction

Mixed Numbers

The root of the word *rational* is *ratio*. Describe the relationship between rational numbers and ratios. _____

_____

_____

 Essential Question

WHY is it helpful to write numbers in different ways?

 Vocab abc **Vocabulary**

rational number
repeating decimal
terminating decimal

 CCSS **Common Core State Standards**

Content Standards
8.NS.1
MP Mathematical Practices
1, 3, 4, 6, 7, 8

Real-World Link

During a recent regular season, a Texas Ranger baseball player had 126 hits and was at bat 399 times. Write a fraction in simplest form to represent the ratio of the number of hits to the number of at bats.

Which MP Mathematical Practices did you use?
Shade the circle(s) that applies.

① Persevere with Problems   ⑤ Use Math Tools

② Reason Abstractly   ⑥ Attend to Precision

③ Construct an Argument   ⑦ Make Use of Structure

④ Model with Mathematics   ⑧ Use Repeated Reasoning

Rational Numbers

Words	A rational number is a number that can be written as the ratio of two integers in which the denominator is not zero.
Symbols	$\frac{a}{b}$, where a and b are integers and $b \neq 0$
Model	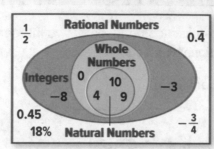

Work Zone

Bar Notation

Bar notation is often used to indicate that a digit or group of digits repeats. The bar is placed above the repeating part. To write 8.636363... in bar notation, write $8.\overline{63}$, not $8.\overline{6}$ or $8.\overline{636}$. To write 0.3444... in bar notation, write $0.3\overline{4}$, not $0.\overline{34}$.

Every rational number can be expressed as a decimal by dividing the numerator by the denominator. The decimal form of a rational number is called a **repeating decimal**. If the repeating digit is zero, then the decimal is a **terminating decimal**.

Rational Number	Repeating Decimal	Terminating Decimal
$\frac{1}{2}$	0.5000...	0.5
$\frac{2}{5}$	0.400...	0.4
$\frac{5}{6}$	0.833...	does not terminate

Examples

Tutor

Write each fraction or mixed number as a decimal.

1. $\frac{5}{8}$

$\frac{5}{8}$ means $5 \div 8$.

$$\begin{array}{r} 0.625 \\ 8\overline{)5.000} \\ -48 \\ \hline 20 \\ -16 \\ \hline 40 \\ -40 \\ \hline 0 \end{array}$$

Divide 5 by 8.

2. $-1\frac{2}{3}$

$-1\frac{2}{3}$ can be rewritten as $\frac{-5}{3}$.

Divide 5 by 3 and add a negative sign.

The mixed number $-1\frac{2}{3}$ can be written as $-1.\overline{6}$.

$$\begin{array}{r} 1.6... \\ 3\overline{)5.0} \\ -3 \\ \hline 20 \\ -18 \\ \hline 2 \end{array}$$

Show your work.

a. _____

b. _____

c. _____

d. _____

Got it? Do these problems to find out.

a. $\frac{3}{4}$ b. $-\frac{2}{9}$

c. $4\frac{13}{25}$ d. $3\frac{1}{11}$

Example

3. In a recent season, St. Louis Cardinals first baseman Albert Pujols had 175 hits in 530 at bats. To the nearest thousandth, find his batting average.

To find his batting average, divide the number of hits, 175, by the number of at bats, 530.

175 ÷ 530 [ENTER] 0.3301886792

Look at the digit to the right of the thousandths place. Since 1 < 5, round down.

Albert Pujols's batting average was 0.330.

Got it? Do this problem to find out.

e. In a recent season, NASCAR driver Jimmie Johnson won 6 of the 36 total races held. To the nearest thousandth, find the part of races he won.

Show your work.

e. _____

Examples

Tutor

4. Write 0.45 as a fraction.

$$0.45 = \frac{45}{100} \quad \text{0.45 is 45 hundredths.}$$

$$= \frac{9}{20} \quad \text{Simplify.}$$

---

5. Write $0.\overline{5}$ as a fraction in simplest form.

Assign a variable to the value $0.\overline{5}$. Let $N = 0.555...$. Then perform operations on N to determine its fractional value.

$$N = 0.555...$$

$$10(N) = 10(0.555...) \quad \text{Multiply each side by 10 because 1 digit repeats.}$$

$$10N = 5.555... \quad \text{Multiplying by 10 moves the decimal point 1 place to the right.}$$

$$-N = 0.555... \quad \text{Subtract } N = 0.555... \text{ to eliminate the repeating part.}$$

$$9N = 5 \quad \text{Simplify.}$$

$$N = \frac{5}{9} \quad \text{Divide each side by 9.}$$

The decimal $0.\overline{5}$ can be written as $\frac{5}{9}$.

6. Write $2.\overline{18}$ as a mixed number in simplest form.

Assign a variable to the value $2.\overline{18}$. Let $N = 2.181818...$. Then perform operations on N to determine its fractional value.

$N = 2.181818...$

$100(N) = 100(2.181818...)$ Multiply each side by 100 because 2 digits repeat.

$100N = 218.181818$ Multiplying by 100 moves the decimal point 2 places to the right.

$\underline{-N = \quad 2.181818...}$ Subtract $N = 2.181818...$ to eliminate the repeating part.

$99N = 216$ Simplify.

$N = \dfrac{216}{99}$ or $2\dfrac{2}{11}$ Divide each side by 99. Simplify.

The decimal $2.\overline{18}$ can be written as $2\dfrac{2}{11}$.

 Show your work.

Got it? Do these problems to find out.

f. _____

g. _____

Write each decimal as a fraction or mixed number in simplest form.

   f. -0.14 **g.** $0.\overline{27}$

Guided Practice

 Check ✓

Write each fraction or mixed number as a decimal. (Examples 1 and 2)

1. $\dfrac{9}{16} =$ _____

2. $-1\dfrac{29}{40} =$ _____

3. $4\dfrac{5}{6} =$ _____

4. Monica won 7 of the 16 science competitions she entered. To the nearest thousandth, find her winning average. (Example 3) _____

Write each decimal as a fraction or mixed number in simplest form. (Examples 4–6)

5. $0.32 =$ _____

6. $-0.\overline{7} =$ _____

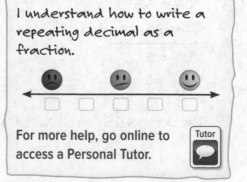

Rate Yourself!

I understand how to write a repeating decimal as a fraction.

For more help, go online to access a Personal Tutor.

7. **Building on the Essential Question** How can you determine if a number is a rational number?

_____

_____

Independent Practice

Go online for Step-by-Step Solutions

Write each fraction or mixed number as a decimal. (Examples 1 and 2)

1. $\frac{2}{5} =$ _____

2. $2\frac{1}{8} =$ _____

3. $\frac{33}{40} =$ _____

4. $\frac{4}{33} =$ _____

5 $-\frac{6}{11} =$ _____

6. $-7\frac{8}{45} =$ _____

7. **MP Identify Repeated Reasoning** The table shows statistics about the students at Carter Junior High. (Example 3)

Number of Siblings	Fraction of Students
None	$\frac{1}{15}$
One	$\frac{1}{3}$
Two	$\frac{5}{12}$
Three	$\frac{1}{6}$
Four or more	$\frac{1}{60}$

   a. Express the fraction of students with no siblings as a decimal.

   _____

   b. Find the decimal equivalent for the fraction of students with three

   siblings. _____

   c. Write the fraction of students with one sibling as a decimal. Round to

   the nearest thousandth. _____

   d. Write the fraction of students with two siblings as a decimal. Round to

   the nearest thousandth. _____

Write each decimal as a fraction or mixed number in simplest form.
(Examples 4–6)

8. $-0.4 =$ _____

9 $-7.32 =$ _____

10. $0.\overline{2} =$ _____

Copy and Solve Write each decimal as a fraction or mixed number in simplest form. Show your work on a separate piece of paper. (Examples 4–6)

11. $-0.\overline{45}$ **12.** $2.\overline{7}$ **13.** 5.55

MP Be Precise Write the length of each insect as a fraction or mixed number and as a decimal.

14.

15.

H.O.T. Problems Higher Order Thinking

16. MP Identify Structure Give an example of a repeating decimal where two digits repeat. Explain why your number is a rational number.

_____

_____

17. MP Persevere with Problems Explain why any rational number is either a terminating or repeating decimal. _____

_____

_____

_____

_____

18. MP Make a Conjecture Compare 0.1 and $0.\overline{1}$, 0.13 and $0.\overline{13}$, and 0.157 and $0.\overline{157}$ when written as fractions. Make a conjecture about expressing repeating decimals like these as fractions.

_____

_____

_____

19. MP Model with Mathematics Write two decimals, one repeating and one terminating, with values between 0 and 1. Then write an inequality that shows the relationship between your two decimals.

_____

_____

Extra Practice

20. Write $\frac{5}{9}$ as a decimal. $0.\overline{5}$ _____

$$
\begin{array}{r}
0.55 \\
9\overline{)5.00} \\
-45 \\
\hline
50 \\
-45 \\
\hline
5,...
\end{array}
$$

Homework
Help

21. Write $7.\overline{15}$ as a mixed number in simplest form. $7\frac{5}{33}$ _____

$$
\begin{aligned}
N &= 7.151515... \\
100(N) &= 100(7.151515...) \\
100N &= 715.151515... \\
-N &= 7.151515... \\
\hline
99N &= 708 \\
N &= \frac{708}{99} \text{ or } 7\frac{5}{33}
\end{aligned}
$$

MP **Identify Repeated Reasoning** Write each fraction or mixed number as a decimal.

22. $\frac{4}{5} =$ _____

23. $5\frac{5}{16} =$ _____

24. $-6\frac{13}{15} =$ _____

Write each decimal as a fraction or mixed number in simplest form.

25. $-1.55 =$ _____

26. $3.\overline{8} =$ _____

27. $-0.\overline{09} =$ _____

Write the rainfall amount for each day as a fraction or mixed number.

28. Friday _____

29. Saturday _____

30. Sunday _____

Day	Rainfall (in.)
Friday	0.08
Saturday	2.4
Sunday	0.035

31. The table shows three popular flavors according to the results of a survey. What is the decimal value of those who liked vanilla, chocolate, or strawberry? Round to the nearest hundredth. _____

Flavor	Fraction
Vanilla	$\frac{3}{10}$
Chocolate	$\frac{1}{11}$
Strawberry	$\frac{1}{18}$

32. Determine if the number in each situation is rational.

a. The position of a submarine with respect to the surface of the water is −225.4 feet.
☐ Rational   ☐ Not Rational

b. A mechanic uses a wrench that is labeled $\frac{13}{16}$-inch.
☐ Rational   ☐ Not Rational

c. The circumference of a pizza in 16π or 50.2654824574… inches.
☐ Rational   ☐ Not Rational

d. Clarence earned a score of 86.7% on his science test.
☐ Rational   ☐ Not Rational

33. The table shows the number of free throws each player made during the last basketball season. Determine if each statement is true or false.

Player	Free Throws Made	Free Throws Attempted
Felisa	18	20
Morgan	13	24
Yasmine	15	22
Gail	10	14

a. Felisa made $\frac{9}{10}$ of her free throw attempts.
☐ True   ☐ False

b. Morgan made $\frac{7}{12}$ of her free throw attempts.
☐ True   ☐ False

c. Yasmine made $\frac{15}{22}$ of her free throw attempts.
☐ True   ☐ False

d. Gail made $\frac{4}{7}$ of her free throw attempts.
☐ True   ☐ False

Common Core Spiral Review

Fill in each ◯ with >, <, or = to make a true statement. **6.NS.7**

34. $2\frac{7}{8}$ ◯ 2.75

35. $\frac{-1}{3}$ ◯ $\frac{-7}{3}$

36. $\frac{5}{7}$ ◯ $\frac{4}{5}$

37. $3\frac{6}{11}$ ◯ $3.\overline{54}$

38. At the grocery store, Karen was comparing the unit price for two different packages of laundry detergent. One package was $0.0733 per ounce. The other package was $3.64 for 52 ounces. Which package had the lower unit price? Explain. **6.RP.2** _____

_____

Powers and Exponents

Real-World Link

Savings Yogi decided to start saving money by putting a penny in his piggy bank, then doubling the amount he saves each week. Use the questions below to find how much money Yogi will save in 8 weeks.

1. Complete the table below to find the amount Yogi saved each week and the total amount in his piggy bank.

Week	0	1	2	3	4	5	6
Weekly Savings	1¢	2¢					
Total Savings	1¢	3¢					

2. How many 2s are multiplied to find his savings in Week 4? [ ]

Week 5? [ ]

3. How much money will Yogi save in Week 8? _____

4. Continue the table to find when he will have enough to buy

a pair of shoes for $80. _____

Week	7	8	9	10	11	12
Weekly Savings						
Total Savings						

Essential Question

WHY is it helpful to write numbers in different ways?

Vocabulary
power
base
exponent

Common Core State Standards

Content Standards
8.EE.1
MP Mathematical Practices
1, 3, 4, 8

Which MP Mathematical Practices did you use?
Shade the circle(s) that applies.

1. Persevere with Problems
2. Reason Abstractly
3. Construct an Argument
4. Model with Mathematics
5. Use Math Tools
6. Attend to Precision
7. Make Use of Structure
8. Use Repeated Reasoning

Write and Evaluate Powers

A product of repeated factors can be expressed as a **power**, that is, using an exponent and a base.

> The **base** is the common factor.

> The **exponent** tells how many times the base is used as a factor.

$$\overbrace{2 \cdot 2 \cdot 2 \cdot 2}^{\text{4 factors}} = 2^4$$

Powers are read in a certain way.

Read and Write Powers		
Power	Words	Factors
3^1	3 to the first power	3
3^2	3 to the second power or 3 squared	$3 \cdot 3$
3^3	3 to the third power or 3 cubed	$3 \cdot 3 \cdot 3$
3^4	3 to the fourth power or 3 to the fourth	$3 \cdot 3 \cdot 3 \cdot 3$
$\vdots$	$\vdots$	$\vdots$
3^n	3 to the nth power or 3 to the nth	$\underbrace{3 \cdot 3 \cdot 3 \cdot \ldots \cdot 3}_{n \text{ factors}}$

Examples

Tutor

Write each expression using exponents.

1. $(-2) \cdot (-2) \cdot (-2) \cdot 3 \cdot 3 \cdot 3 \cdot 3$

The base -2 is a factor 3 times, and the base 3 is a factor 4 times.

$(-2) \cdot (-2) \cdot (-2) \cdot 3 \cdot 3 \cdot 3 \cdot 3 = (-2)^3 \cdot 3^4$

2. $a \cdot b \cdot b \cdot a \cdot b$

Use the properties of operations to rewrite and group like bases together. The base a is a factor 2 times, and the base b is a factor 3 times.

$$a \cdot b \cdot b \cdot a \cdot b = a \cdot a \cdot b \cdot b \cdot b$$
$$= a^2 \cdot b^3$$

> Show your work.

a. _____

b. _____

c. _____

Got it? Do these problems to find out.

a. $\dfrac{1}{2} \cdot \dfrac{1}{2} \cdot \dfrac{1}{2} \cdot \dfrac{1}{2}$

b. $4 \cdot 4 \cdot 4 \cdot 5 \cdot 5$

c. $m \cdot m \cdot n \cdot n \cdot m$

Example

3. Evaluate $\left(-\dfrac{2}{3}\right)^4$.

$$\left(-\dfrac{2}{3}\right)^4 = \left(-\dfrac{2}{3}\right) \cdot \left(-\dfrac{2}{3}\right) \cdot \left(-\dfrac{2}{3}\right) \cdot \left(-\dfrac{2}{3}\right) \qquad \text{Write the power as a product.}$$

$$= \dfrac{16}{81} \qquad \text{Multiply.}$$

Got it? Do these problems to find out.

d. 4^4 e. $(-2)^6$ f. $\left(\dfrac{1}{5}\right)^3$

Evaluate
Remember that to evaluate an expression means to find its value.

Show your work.

d. _____

e. _____

f. _____

Example

4. The deck of a skateboard has an area of about $2^5 \cdot 7$ square inches. What is the area of the skateboard deck?

$$2^5 \cdot 7 = 2 \cdot 2 \cdot 2 \cdot 2 \cdot 2 \cdot 7 \qquad \text{Write the power as a product.}$$

$$= (2 \cdot 2 \cdot 2 \cdot 2 \cdot 2) \cdot 7 \qquad \text{Associative Property}$$

$$= 32 \cdot 7 \text{ or } 224 \qquad \text{Multiply.}$$

The area of the skateboard deck is about 224 square inches.

Got it? Do this problem to find out.

g. A school basketball court has an area of $2^3 \cdot 3 \cdot 5^2 \cdot 7$ square feet. What is the area of a school basketball court?

g. _____

Examples

Evaluate each expression if $a = 3$ and $b = 5$.

5. $a^2 + b^4$

$$a^2 + b^4 = 3^2 + 5^4 \qquad \text{Replace } a \text{ with 3 and } b \text{ with 5.}$$

$$= (3 \cdot 3) + (5 \cdot 5 \cdot 5 \cdot 5) \qquad \text{Write the powers as products.}$$

$$= 9 + 625 \text{ or } 634 \qquad \text{Add.}$$

6. $(a - b)^2$

$$(a - b)^2 = (3 - 5)^2 \qquad \text{Replace } a \text{ with 3 and } b \text{ with 5.}$$

$$= (-2)^2 \qquad \text{Perform operations in the parentheses first.}$$

$$= (-2) \cdot (-2) \text{ or } 4 \qquad \text{Write the powers as products. Then simplify.}$$

h. _____

i. _____

j. _____

Got it? Do these problems to find out.

Evaluate each expression if $c = -4$ and $d = 9$.

    h. $c^3 + d^2$ **i.** $(c + d)^3$ **j.** $d^3 - (c^2 - 2)$

Guided Practice

Write each expression using exponents. (Examples 1 and 2)

1. $(-11)(-11)(-11) = $ _____

2. $2 \cdot 2 \cdot 2 \cdot 3 \cdot 3 \cdot 3 = $ _____

3. $r \cdot s \cdot r \cdot r \cdot s \cdot s \cdot r \cdot r = $ _____

Evaluate each expression. (Example 3)

4. $2^6 = $ _____

5. $(-4)^4 = $ _____

6. $\left(\dfrac{1}{7}\right)^3 = $ _____

7. The table shows the average weights of some endangered mammals. What is the weight of each animal? (Example 4)

_____

Animal	Weight (lb)
Black bear	$2 \cdot 5^2 \cdot 7$
Key deer	$3 \cdot 5^2$
Panther	$2^3 \cdot 3 \cdot 5$

Evaluate each expression if $x = 2$ and $y = 10$. (Examples 5 and 6)

8. $x^2 + y^4 = $ _____

9. $(x^2 + y)^3 = $ _____

10. 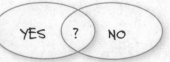 **Building on the Essential Question** How can I write repeated multiplication using powers? _____

_____

_____

Rate Yourself!

Are you ready to move on?
Shade the section that applies.

    YES   ?   NO

For more help, go online to access a Personal Tutor.

Name ______________________ My Homework ______________________

Extra Practice

17. Write $3 \cdot p \cdot p \cdot p \cdot 3 \cdot 3$ using exponents.

$3^3 \cdot p^3$

$3 \cdot p \cdot p \cdot p \cdot 3 \cdot 3 = 3 \cdot 3 \cdot 3 \cdot p \cdot p \cdot p$

$= 3^3 \cdot p^3$

Homework Help

18. Evaluate $x^3 + y^4$ if $x = -3$ and $y = 4$.

229

$x^3 + y^4 = (-3)^3 + 4^4$

$= (-3) \cdot (-3$ … $) + 4 \cdot 4 \cdot 4 \cdot 4$

$= (-27)$ … 256

$= 229$

Write each expression using exponents.

19. $\left(-\frac{5}{6}\right)\left(-\frac{5}{6}\right)\left(-\frac{5}{6}\right) =$ ______

20. $s \cdot (7) \cdot s \cdot (7) \cdot (7) =$ ______

21. $4 \cdot b \cdot b \cdot 4 \cdot b \cdot b =$ ______

Evaluate each expression.

22. $k^4 \cdot m$, if $k = 3$ and $m = \frac{5}{6}$

23. $(c^3 + d^4)^2 - (c + d)^3$, if $c = -1$ and $d = 2$

Fill in each ◯ with <, >, or = to make a true statement.

24. $(6 - 2)^2 + 3 \cdot 4$ ◯ 5^2

25. $5 + 7^2 + 3^3$ ◯ 3^4

26. $\left(\frac{1}{2}\right)^4$ ◯ $\left(\frac{1}{4}\right)^2$

27. MP **Multiple Representations** A square has a side length of *s* inches.

a. **Tables** Copy and complete the table showing the side length, perimeter, and area of the square on a separate piece of paper.

b. **Graphs** On a sepa… …iece of grid paper, graph the ordered pairs (side … perimeter) and (side length, area) on the same co… …ate plane. Then connect the points for each set.

c. **Words** On a separate s… …of paper, compare and contrast the graphs of t… …meter and area of the square. Which graph is … .

Side Length (in.)	Perimeter (in.)	Area (in^2)
1	4	1
2		
3		
4		
5		
⋮		
10		

28. Hard drive storage capacity is measured in bytes using the metric system. The metric system is based on powers of 10. For example, 1 kilobyte is equal to 1,000 bytes or 103 bytes. The table shows some common units of storage capacity. Select the correct power of 10 to complete the table.

Unit	Number of Bytes
megabyte	1,000,000
terabyte	1,000,000,000,000
gigabyte	1,000,000,000

Unit	Power of 10
megabyte	
terabyte	
gigabyte	

| 10^3 | 10^6 | 10^9 | 10^{12} | 10^{15} |

29. A cube has the dimensions shown below.

6 in.

What is the volume of the cube expressed as a power?

(CCSS) Common Core Spiral Review

30. The table below shows the number of ants in an ant farm on different days. The number of ants doubles every ten days. **7.EE.3**

Day	51	61	71
Number of Ants	320	640	1,280

a. How many ants were in the farm on Day 1? _____

b. How many ants will be in the farm on Day 91? _____

Add. 7.NS.1

31. $-12 + (-19) =$ _____

32. $-8 + (-11) =$ _____

33. $-5 + 6 =$ _____

Multiply and Divide Monomials

Real-World Link

Arachnids Spiders in North America can range in size from 1 millimeter in length to 7.6 centimeters in length. Use the table to see how other metric measurements of length are related to the millimeter.

Unit of Length	Times Longer than a Millimeter	Written Using Powers
Millimeter	1	10^0
Centimeter	$1 \times 10 = \boxed{}$	10^1
Decimeter	$10 \times 10 = \boxed{}$	$10^1 \times 10^1 = 10^2$
Meter	$100 \times 10 = 1{,}000$	$10^2 \times 10^1 = 10^{\boxed{}}$
Dekameter	$1{,}000 \times 10 = 10{,}000$	$10^3 \times 10^1 = 10^{\boxed{}}$
Hectometer	$10{,}000 \times 10 = \boxed{}$	$10^4 \times 10^1 = 10^5$
Kilometer	$100{,}000 \times 10 = \boxed{}$	$10^5 \times 10^1 = 10^{\boxed{}}$

Essential Question

WHY is it helpful to write numbers in different ways?

Vocabulary

monomial

Common Core State Standards

Content Standards
8.EE.1

MP Mathematical Practices
1, 3, 4, 7

1. Look at the entries in the last column. What do you observe about the exponents of the factors and the exponent of the product for each entry? _____

   _____

2. A *megameter* is 100,000,000 × 10 or 1,000,000,000 times longer than a millimeter. Extend the pattern to write this number using powers. _____

Which MP Mathematical Practices did you use?
Shade the circle(s) that applies.

① Persevere with Problems

② Reason Abstractly

③ Construct an Argument

④ Model with Mathematics

⑤ Use Math Tools

⑥ Attend to Precision

⑦ Make Use of Structure

⑧ Use Repeated Reasoning

Product of Powers

Words To multiply powers with the same base, add their exponents.

Examples Numbers
$$2^4 \cdot 2^3 = 2^{4+3} \text{ or } 2^7$$

Algebra
$$a^m \cdot a^n = a^{m+n}$$

A **monomial** is a number, a variable, or a product of a number and one or more variables. You can use the Laws of Exponents to simplify monomials.

$$\overbrace{2 \text{ factors}}^{} \quad \overbrace{4 \text{ factors}}^{}$$
$$3^2 \cdot 3^4 = \underbrace{(3 \cdot 3) \cdot (3 \cdot 3 \cdot 3 \cdot 3)}_{6 \text{ factors}} \text{ or } 3^6$$

Notice that the sum of the original exponents is the exponent in the final product.

Examples

Tutor

Simplify using the Laws of Exponents.

1. $5^2 \cdot 5$

$5^2 \cdot 5 = 5^2 \cdot 5^1$ $5 = 5^1$ **Check** $5^2 \cdot 5 = (5 \cdot 5) \cdot 5$

$\quad\quad = 5^{2+1}$ The common base is 5. $= 5 \cdot 5 \cdot 5$

$\quad\quad = 5^3 \text{ or } 125$ Add the exponents. Simplify. $= 5^3 \checkmark$

2. $c^3 \cdot c^5$

$c^3 \cdot c^5 = c^{3+5}$ The common base is c.

$\quad\quad = c^8$ Add the exponents.

3. $-3x^2 \cdot 4x^5$

Show your work.

$-3x^2 \cdot 4x^5 = (-3 \cdot 4)(x^2 \cdot x^5)$ Commutative and Associative Properties

$\quad\quad = (-12)(x^{2+5})$ The common base is x.

$\quad\quad = -12x^7$ Add the exponents.

Got it? Do these problems to find out.

a. _____

b. _____

c. _____

a. $9^3 \cdot 9^2$ b. $a^3 \cdot a^2$ c. $-2m(-8m^5)$

Quotient of Powers

Words   To divide powers with the same base, subtract their exponents.

Examples   Numbers

$$\frac{3^7}{3^3} = 3^{7-3} \text{ or } 3^4$$

Algebra

$$\frac{a^m}{a^n} = a^{m-n}, \text{ where } a \neq 0$$

 STOP and Reflect

Explain below why the Quotient of Powers rule cannot be used to simplify the expression $\frac{x^5}{y^3}$.

There is also a Law of Exponents for dividing powers with the same base.

$$\frac{5^7}{5^4} = \frac{\overbrace{5 \cdot 5 \cdot 5 \cdot \cancel{5} \cdot \cancel{5} \cdot \cancel{5} \cdot \cancel{5}}^{\text{7 factors}}}{\underbrace{\cancel{5} \cdot \cancel{5} \cdot \cancel{5} \cdot \cancel{5}}_{\text{4 factors}}} \text{ or } 5^3$$

Notice that the difference of the original exponents is the exponent in the final quotient.

Examples

 Tutor

Simplify using the Laws of Exponents.

4. $\dfrac{4^8}{4^2}$

$\dfrac{4^8}{4^2} = 4^{8-2}$ The common base is 4.

   $= 4^6$ or 4,096   Simplify.

5. $\dfrac{n^9}{n^4}$

$\dfrac{n^9}{n^4} = n^{9-4}$ The common base is n.

   $= n^5$ Simplify.

6. $\dfrac{2^5 \cdot 3^5 \cdot 5^2}{2^2 \cdot 3^4 \cdot 5}$

$\dfrac{2^5 \cdot 3^5 \cdot 5^2}{2^2 \cdot 3^4 \cdot 5} = \left(\dfrac{2^5}{2^2}\right)\left(\dfrac{3^5}{3^4}\right)\left(\dfrac{5^2}{5}\right)$ Group by common base.

   $= 2^3 \cdot 3^1 \cdot 5^1$ Subtract the exponents.

   $= 8 \cdot 3 \cdot 5$ $2^3 = 8$

   $= 120$ Simplify.

Got it? Do these problems to find out.

d. $\dfrac{5^7}{5^4}$

e. $\dfrac{x^{10}}{x^3}$

f. $\dfrac{12w^5}{2w}$

g. $\dfrac{3^4 \cdot 5^2 \cdot 7^5}{3^2 \cdot 5 \cdot 7^3}$

h. $\dfrac{5^6 \cdot 7^4 \cdot 8^3}{5^4 \cdot 7^2 \cdot 8^2}$

i. $\dfrac{(-2)^5 \cdot 3^4 \cdot 5^7}{(-2)^2 \cdot 3 \cdot 5^4}$

 Show your work.

d. _____

e. _____

f. _____

g. _____

h. _____

i. _____

Example

7. Hawaii's total shoreline is about 2^{10} miles long. New Hampshire's shoreline is about 2^7 miles long. About how many times longer is Hawaii's shoreline than New Hampshire's?

To find how many times longer, divide 2^{10} by 2^7.

$$\frac{2^{10}}{2^7} = 2^{10-7} \text{ or } 2^3 \qquad \text{Quotient of Powers}$$

Hawaii's shoreline is about 2^3 or 8 times longer.

Guided Practice

Simplify using the Laws of Exponents. (Examples 1–6)

1. $4^5 \cdot 4^3 = $ _____

2. $-2a(3a^4) = $ _____

3. $\dfrac{y^8}{y^5} = $ _____

4. $\dfrac{24k^9}{6k^6} = $ _____

5. $\dfrac{2^2 \cdot 3^3 \cdot 4^5}{2 \cdot 3 \cdot 4^4} = $ _____

6. $\dfrac{(-3)^4 \cdot (-4)^3 \cdot 5^2}{(-3)^2 \cdot (-4) \cdot 5} = $ _____

7. The table shows the number of people worldwide that speak certain languages. How many times as many people speak French than Sicilian?

(Example 7) _____

Language	Total (millions)
French	2^6
Sicilian	2^2

8. **Building on the Essential Question** How can I use the properties of integer exponents to simplify algebraic and numeric expressions? _____

_____

_____

_____

Rate Yourself!

Are you ready to move on? Shade the section that applies.

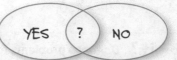

YES   ?   NO

For more help, go online to access a Personal Tutor.

FOLDABLES Time to update your Foldable!

Independent Practice

Go online for Step-by-Step Solutions eHelp

Simplify using the Laws of Exponents. (Examples 1–6)

1. $(-6)^2 \cdot (-6)^5 =$ _____

2. $-4a^5(6a^5) =$ _____

3. $(-7a^4bc^3)(5ab^4c^2) =$ _____

4. $\dfrac{8^{15}}{8^{13}} =$ _____

5. $\dfrac{16t^4}{8t} =$ _____

6. $\dfrac{x^6y^{14}}{x^4y^9} =$ _____

7. $\dfrac{3^4x^4}{3x^2} =$ _____

8. $\dfrac{4^5 \cdot 5^3 \cdot 6^2}{4^4 \cdot 5^2 \cdot 6} =$ _____

9. $\dfrac{6^3 \cdot 6^6 \cdot 6^4}{6^2 \cdot 6^3 \cdot 6^3} =$ _____

10. $\dfrac{(-2)^5 \cdot (-3)^4 \cdot (-5)^3}{(-2)^3 \cdot (-3) \cdot (-5)^2} =$

_____

_____

11. The processing speed of a certain computer is 10^{11} instructions per second. Another computer has a processing speed that is 10^3 times as fast. How many instructions per second can the faster computer process?
(Example 7)

_____

12. The table shows the seating capacity of two different facilities. About how many times as great is the capacity of Madison Square Garden in New York than a typical movie theater? (Example 7)

_____

Place	Seating Capacity
Movie theater	3^5
Madison Square Garden	3^9

13. Refer to the information in the table.

a. How many times as great is one quadrillion than one million?

_____

b. One quintillion is one trillion times as great as what number?

_____

Power of Ten	U.S. Name
10^3	one thousand
10^6	one million
10^9	one billion
10^{12}	one trillion
10^{15}	one quadrillion
10^{18}	one quintillion

MP Persevere with Problems Find each missing exponent.

14. $(6^{\bullet})(6^3) = 6^5$ _____

15. $3x^{\bullet} \cdot 4x^3 = 12x^{12}$ _____

16. $p^3 \cdot p^{\bullet} \cdot p^2 = p^9$ _____

17. $\dfrac{3^{\bullet}}{3^2} = 3^4$ _____

18. $\dfrac{5^9}{5^{\bullet}} = 5^4$ _____

19. $2x^{\bullet} \cdot \dfrac{3x^2}{x^6} = 6x^3$ _____

H.O.T. Problems Higher Order Thinking

20. MP Identify Structure Write a multiplication expression with a product of 5^{13}.

_____

21. MP Justify Conclusions Is $\dfrac{3^{100}}{3^{99}}$ greater than, less than, or equal to 3?

Explain your reasoning to a classmate. _____

_____

22. MP Persevere with Problems What is twice 2^{30}? Write using exponents. Explain your reasoning.

_____

_____

23. MP Use a Counterexample Determine whether the statement below is true or false. If *true*, explain your reasoning. If *false*, give a counterexample.

For any integer a, $(-a)^2 = -a^2$.

_____

_____

Extra Practice

Simplify using the Laws of Exponents.

24. $(3x^8)(5x) = \underline{15x^9}$

$(3x^8)(5x) = 3 \cdot 5 \cdot x^8 \cdot x$

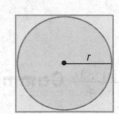 Homework Help ➡

$= 15 \cdot x^{8+1}$

$= 15x^9$

25. $\dfrac{h^7}{h^6} = \underline{h^1 \text{ or } h}$

$\dfrac{h^7}{h^6} = h^{7-6}$

$= h^1 \text{ or } h$

26. $2g^2 \cdot 7g^6 = $ _____

27. $(8w^4)(-w^7) = $ _____

28. $(-p)(-9p^2) = $ _____

29. $\dfrac{2^9}{2} = $ _____

30. $\dfrac{36d^{10}}{6d^5} = $ _____

31. $\dfrac{5^3 \cdot 7^5 \cdot 10}{5 \cdot 7^4} = $ _____

32. $\dfrac{(-3)^2 \cdot 4^3 \cdot (-1)^8}{4 \cdot (-1)^5} = $ _____

33. **MP Persevere with Problems** The figure at the right is composed of a circle and a square. The circle touches the square at the midpoints of the four sides.

a. What is the length of one side of the square? _____

b. The formula $A = \pi r^2$ is used to find the area of a circle. The formula $A = 4r^2$ can be used to find the area of the square. Write the ratio of the area of the circle to the area of the square in simplest form.

_____

c. Complete the table.

Radius (units)	2	3	4	2r
Area of Circle (units²)	$\pi(2)^2$ or 4π			
Length of 1 Side of the Square	4			
Area of Square (units²)	4^2 or 16			
Ratio $\dfrac{\text{(Area of circle}}{\text{Area of square)}}$				

d. What can you conclude about the relationship between the areas of the circle and the square? _____

34. Which expression(s) could be used to represent the area of the rectangle? Select all that apply.

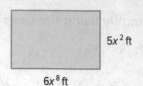

$5x^2$ ft

$6x^8$ ft

☐ $\dfrac{6x^8}{5x^2}$ ft^2 ☐ $6x^8(5x^2)$ ft^2

☐ $30x^{10}$ ft^2 ☐ $\dfrac{6}{5}x^6$ ft^2

35. The table shows the approximate populations of four states.

State	Alabama	Idaho	Illinois	Wyoming
Approximate Population	3^{14}	3^{13}	3^{15}	3^{12}

Select the correct state to make each statement true.

Idaho
Alabama
Illinois
Wyoming

Statement 1: The population of ⬚ is about $\frac{1}{3}$ the population of Alabama.

Statement 2: The population of ⬚ is about 9 times greater than the population of Wyoming.

Statement 3: The population of ⬚ is about 27 times

greater than the population of ⬚ .

 Common Core Spiral Review

Multiply or divide. 7.NS.2

36. $14(-2) = $ _____

37. $-20(-3) = $ _____

38. $-5(7) = $ _____

39. $-12 \div (-4) = $ _____

40. $63 \div (-7) = $ _____

41. $250 \div (-50) = $ _____

42. Three-fourths of a pan of lasagna is to be divided equally among 6 people. What part of the lasagna will each person receive? 6.NS.1

_____

Powers of Monomials

Real-World Link

Aquariums The Marine Club at Westview Middle School purchased an aquarium. The aquarium is in the shape of a cube with a side length of 2^4 inches. Use the questions to find the amount of water the aquarium willl hold.

1. Write a multiplication expression to represent the volume of the aquarium. _____

2. Simplify the expression. Write as a single power of 2. ⬚

3. Using 2^4 as the base, write the multiplication expression $2^4 \cdot 2^4 \cdot 2^4$ using an exponent. ⬚

4. Explain why $(2^4)^3 = 2^{12}$. _____

5. Use a calculator to find the volume of the tank.

   ⬚ cubic inches

6. One gallon of water is equal to 231 cubic inches. Write an expression to find how many gallons of water the tank will hold if

   it is filled to the top. $\dfrac{⬚}{⬚}$

7. How many gallons of water will the aquarium hold? Round your answer to the nearest gallon. ⬚ gallons

Essential Question

WHY is it helpful to write numbers in different ways?

Common Core State Standards

Content Standards
8.EE.1

MP Mathematical Practices
1, 3, 4, 7

Which MP **Mathematical Practices** did you use?
Shade the circle(s) that applies.

① Persevere with Problems

② Reason Abstractly

③ Construct an Argument

④ Model with Mathematics

⑤ Use Math Tools

⑥ Attend to Precision

⑦ Make Use of Structure

⑧ Use Repeated Reasoning

Power of a Power

Words   To find the power of a power, multiply the exponents.

Examples   Numbers   Algebra
$(5^2)^3 = 5^{2 \cdot 3}$ or 5^6 $(a^m)^n = a^{m \cdot n}$

You can use the rule for finding the *product* of powers to discover another Law of Exponents for finding the *power* of a power.

$$\overbrace{(6^4)^5 = (6^4)(6^4)(6^4)(6^4)(6^4)}^{\text{5 factors}}$$

$$= 6^{4+4+4+4+4} \quad \text{Apply the rule for the product of powers.}$$

$$= 6^{20}$$

Notice that the product of the original exponents, 4 and 5, is the final power 20.

Examples

Simplify using the Laws of Exponents.

1. $(8^4)^3$

$(8^4)^3 = 8^{4 \cdot 3}$ Power of a Power

$\quad\quad = 8^{12}$ Simplify.

 Show your work.

2. $(k^7)^5$

$(k^7)^5 = k^{7 \cdot 5}$ Power of a Power

$\quad\quad = k^{35}$ Simplify.

a. _____

b. _____

Got it?   Do these problems to find out.

a. $(2^5)^2$ b. $(w^4)^6$ c. $[(3^2)^3]^2$

c. _____

Power of a Product

Words To find the power of a product, find the power of each factor and multiply.

Examples Numbers | Algebra

$(6x^2)^3 = (6)^3 \cdot (x^2)^3$ or $216x^6$ $(ab)^m = a^m b^m$

Extend the power of a *power* rule to find the Laws of Exponents for the power of a *product*.

$$\overset{\text{5 factors}}{\overbrace{(3a^2)^5 = (3a^2)(3a^2)(3a^2)(3a^2)(3a^2)}}$$

$$= 3 \cdot 3 \cdot 3 \cdot 3 \cdot 3 \cdot a^2 \cdot a^2 \cdot a^2 \cdot a^2 \cdot a^2$$

$$= 3^5 \cdot (a^2)^5 \qquad \text{Write using powers.}$$

$$= 243 \cdot a^{10} \text{ or } 243a^{10} \qquad \text{Power of a Power}$$

> **Common Error**
> When finding the power of a power, do not add the exponents.
> $(8^4)^3 = 8^{12}$, not 8^7.

Examples

Simplify using the Laws of Exponents.

3. $(4p^3)^4$

$(4p^3)^4 = 4^4 \cdot p^{3 \cdot 4} \qquad$ Power of a Product

$\qquad\quad = 256p^{12} \qquad$ Simplify.

. . . .

4. $(-2m^7n^6)^5$

$(-2m^7n^6)^5 = (-2)^5 m^{7 \cdot 5} n^{6 \cdot 5} \qquad$ Power of a Product

$\qquad\qquad\quad = -32m^{35}n^{30} \qquad$ Simplify.

> **Got it?** Do these problems to find out.

d. $(8b^9)^2$ e. $(6x^5y^{11})^4$ f. $(-5w^2z^8)^3$

Show your work.

d. _____

e. _____

f. _____

Example

5. A magazine offers a special service to its subscribers. If they scan the square logo shown on a smartphone, they can receive special offers from the magazine. Find the area of the logo.

$A = s^2$ Area of a square

$A = (7a^4b)^2$ Replace s with $7a^4b$.

$A = 7^2(a^4)^2(b^1)^2$ Power of a Product

$A = 49a^8b^2$ Simplify.

$\longleftarrow 7a^4b \longrightarrow$

The area of the logo is $49a^8b^2$ square units.

Guided Practice

Check ✓

Simplify using the Laws of Exponents. (Examples 1–4)

1. $(3^2)^5 =$ _____

Show your work.

2. $(h^6)^4 =$ _____

3. $[(2^3)^2]^3 =$ _____

4. $(7w^7)^3 =$ _____

5. $(5g^8k^{12})^4 =$ _____

6. $(-6r^5s^9)^2 =$ _____

7. The floor of the commons room at King Middle School is in the shape of a square with side lengths of x^2y^3 feet. New tile is going to be put on the floor of the room. Find the area of the floor. (Example 5) _____

8. **Building on the Essential Question** How does the Product of Powers law apply to finding the power of a power?

_____

_____

_____

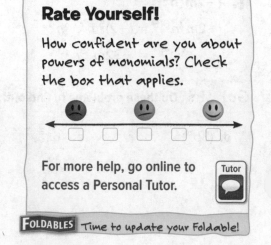

Rate Yourself!

How confident are you about powers of monomials? Check the box that applies.

For more help, go online to access a Personal Tutor.

Tutor

FOLDABLES Time to update your Foldable!

Independent Practice

Go online for Step-by-Step Solutions

Simplify using the Laws of Exponents. (Examples 1–4)

1. $(4^2)^3 =$ _____

2. $(5^3)^3 =$ _____

3. $(d^7)^6 =$ _____

4. $(h^4)^9 =$ _____

5. $[(3^2)^2]^2 =$ _____

6. $[(5^2)^2]^2 =$ _____

7. $(5j^6)^4 =$ _____

8. $(11c^4)^3 =$ _____

9. $(6a^2b^6)^3 =$ _____

10. $(2m^5n^{11})^6 =$ _____

11. $(-3w^3z^8)^5 =$ _____

12. $(-5r^4s^{12})^4 =$ _____

13 A shipping box is in the shape of a cube. Each side measures $3c^6d^2$ inches. Express the volume of the cube as a monomial. (Example 5)

_____

14. Tamara is decorating her patio with a planter in the shape of a cube like the one shown. Find the volume of the planter. (Example 5)

_____

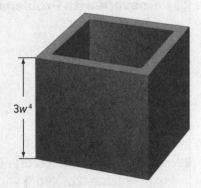

$3w^4$

Copy and Solve **Simplify. Show your work on a separate sheet of paper.**

15. $[(3x^2y^3)^2]^3$

16. $\left(\frac{3}{5}a^6b^9\right)^2$

17 $(-2v^7)^3 (-4v^2)^4$

18. **MP Identify Structure** Draw a line connecting the Law(s) of Exponents you would use to simplify each of the expressions. Then simplify each one.

$(a^9)^3 =$ _____

Product of Powers

$(m^8) \div (m^4) =$ _____

Quotient of Powers

$5x^2 \cdot (-7x^4) =$ _____

Power of a Power

$\dfrac{(xy^4)^3}{xy} =$ _____

Power of a Product

$(n^6)^8 =$ _____

H.O.T. Problems Higher Order Thinking

19. **MP Reason Inductively** The table gives the area and volume of a square and cube, respectively, with side lengths shown.

a. Complete the table.

b. Describe how the area and volume are each affected if the side length is doubled. Then describe how they are each affected if the side length is tripled.

Side Length (units)	x	$2x$	$3x$
Area of Square (units²)	x^2		
Volume of Cube (units³)	x^3		

_____

_____

_____

MP Persevere with Problems Solve each equation for x.

20. $(7^x)^3 = 7^{15}$ _____

21. $(-2m^3n^4)^x = -8m^9n^{12}$ _____

22. **MP Reason Inductively** Compare how you would correctly simplify the expressions $(2a^3)(4a^6)$ and $(2a^3)^6$.

_____

_____

_____

Extra Practice

Simplify using the Laws of Exponents.

23. $(2^2)^7 = $ ___2^{14}___

$$(2^2)^7 = 2^{2 \cdot 7}$$
$$= 2^{14}$$

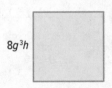 Homework Help

24. $(8v^9)^5 = $ ___32,768v^{45}___

$$(8v^9)^5 = 8^5 \cdot v^{9 \cdot 5}$$
$$= 32,768v^{45}$$

25. $(3^4)^2 = $ _____

26. $(m^8)^5 = $ _____

27. $(z^{11})^5 = $ _____

28. $[(4^3)^2]^2 = $ _____

29. $[(2^3)^3]^2 = $ _____

30. $(14y)^4 = $ _____

Express the area of each square as a monomial.

31. _____

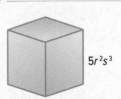

$8g^3h$

32. _____

$12d^6e^7$

Express the volume of each cube as a monomial.

33. _____

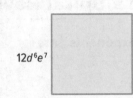

$5r^2s^3$

34. _____

$7m^6n^9$

Simplify.

35. $(0.5k^5)^2 = $ _____

36. $(0.3p^7)^3 = $ _____

37. $\left(\frac{1}{4}w^5z^3\right)^2 = $ _____

38. **MP Persevere with Problems** A ball is dropped from the top of a building. The expression $4.9x^2$ gives the distance in meters the ball has fallen after x seconds. Write and simplify an expression that gives the distance in meters the ball has fallen after x^2 seconds. after x^3 seconds.

39. Manny has four pieces of carpet in the shape of a square like the one shown. He wants to use them together to carpet a portion of his basement. What is the area of the space he can cover with the carpet?

$2x^2$ yards

40. Select the correct expression to represent the volume of each cube.

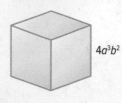 $4a^3b^2$

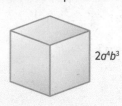 $2a^4b^3$

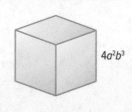 $4a^2b^3$

$2a^{12}b^6$	$8a^{12}b^9$
$4a^6b^3$	$12a^6b^5$
$4a^9b^6$	$64a^6b^9$
$6a^6b^6$	$64a^9b^6$
$8a^{12}b^6$	$64a^9b^{12}$

Common Core Spiral Review

Simplify using the Laws of Exponents. 8.EE.1

41. $6^4 \cdot 6^7 =$ _____

42. $18^3 \cdot 18^5 =$ _____

43. $(-3x^{11})(-6x^3) =$ _____

44. $(-9a^4)(2a^7) =$ _____

45. The table shows the heights of some United States waterfalls. What is the height of each waterfall? 6.EE.1

_____

_____

Waterfall	Height (ft)
Bridalveil (California)	$2^2 \cdot 5 \cdot 31$
Fall Creek (Tennessee)	2^8
Shoshone (Idaho)	$2^2 \cdot 53$

MP Problem-Solving Investigation

The Four-Step Plan

Case #1 Texting Trail

Lillian received a text about a concert. She forwarded the text to two of her friends. They each forwarded it to two more friends, and so on.

How many texts were sent at the 4th stage?

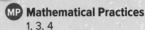

CCSS Content Standards
8.EE.1

MP Mathematical Practices
1, 3, 4

Understand *What are the facts?*

You know that each person at each stage sends a text to two people. You can use counters to represent the trail of texts sent.

Plan *What is your strategy to solve this problem?*

Use red counters to represent the texts in the first stage. Use yellow counters to show the texts sent at the second stage. Continue the pattern. Draw the counters representing the number of texts sent in the 4th stage.

Solve *How can you apply the strategy?*

1st stage
2nd stage
3rd stage
4th stage

There are ☐ counters in the 4th row. So, ☐ texts were sent during the 4th stage.

Check *Does the answer make sense?*

The number of texts at each stage is a power of 2. So, find 2^4.
Since $2^4 = 16$, the answer is correct. ✓

Analyze the Strategy Tools Tutor

MP Justify Conclusions At what stage would there be more than 1,000 texts sent? Explain.

_____

_____

_____

Case #2 Green Mileage

A test of a hybrid car resulted in 4,840 miles driven using 88 gallons of gas.

At this rate, how many gallons of gas will this vehicle need to travel 1,155 miles?

Understand

Read the problem. What are you being asked to find?

I need to find _____.

Underline key words and values in the problem. What information do you know?

The hybrid car can travel _____ miles using _____ gallons of gas.

Is there any information that you do *not* need to know?

I do not need to know _____.

Plan

How do the facts relate to one another?

_____

Solve

Write and solve a proportion comparing miles to gallons. Let *g* represent the amount of gas needed to travel 1,155 miles.

$\dfrac{\text{miles}}{\text{gallons}}$ ⟶ $\dfrac{\boxed{}}{\boxed{}} = \dfrac{\boxed{}}{\boxed{}}$

How many gallons of gas will the car use to travel 1,155 miles? $\boxed{}$

Check

Use information from the problem to check your answer.

_____

_____

Work with a small group to solve the following cases.
Show your work on a separate piece of paper.

Case #3 Class Trip

All of Mr. Bassett's science classes are going to the Natural History Museum. A tour guide is needed for each group of eight students. His classes have 28 students, 35 students, 22 students, 33 students, and 22 students.

How many tour guides are needed?

_____

Case #4 Gardening

Mrs. Lopez is designing her garden in the shape of a rectangle. The area of her garden is 2 times greater than the area of the rectangle shown.

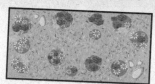

$8s^2$ ft

$4s^3t$ ft

Write the area of Mrs. Lopez's garden in simplest form.

_____

Case #5 Toothpicks

Hector will make the figures at the right using toothpicks.

Figure 1   Figure 2   Figure 3   Figure 4

Write an expression that can be used to find the number of toothpicks needed to make any figure. Then find the number of toothpicks needed to make the 100th figure.

_____

Use any strategy!

Case #6 Number Sense

Study the following sequence:

$1 - \frac{1}{2}, 1 - \frac{1}{2}, 1 - \frac{1}{3}, 1 - \frac{1}{4}, ..., 1 - \frac{1}{48}, 1 - \frac{1}{49}$, and $1 - \frac{1}{50}$

What is the product of all of the terms?

_____

Mid-Chapter Check

Vocabulary Check

1. **MP** **Be Precise** Define *power* using the words *base* and *exponent*. Give an example of a power and label the base and exponent. (Lesson 2)

2. Describe the Product of Powers rule. Give an example. (Lesson 3)

Skills Check and Problem Solving

3. Write $1\frac{7}{16}$ as a decimal. (Lesson 1) _____

4. Write $0.\overline{15}$ as a fraction in simplest form.
   (Lesson 1) _____

5. The mass of a baseball glove is $5 \cdot 5 \cdot 5 \cdot 5$ grams. Write the mass using exponents. Then find the value of the expression. (Lesson 2) _____

Simplify using the Laws of Exponents. (Lessons 3 and 4)

6. $2^3a^7 \cdot 2a^3 =$ _____

7. $\dfrac{24y^4}{4y^2} =$ _____

8. $(2p^3r^2)^3 =$ _____

9. **MP** **Persevere with Problems** Write two algebraic expressions, one with a quotient of x^5 and one with a product of x^5. (Lesson 3)

Lesson 5

Negative Exponents

Real-World Link

Insects The table shows the approximate wing beats per minute for certain insects.

Insect	Wing Beats per Minute
house fly	10,000
small butterfly	100

1. Write a ratio in simplest form that compares the number of wing beats for a butterfly to a housefly. $\dfrac{\square}{\square}$

2. Write the ratio as a fraction with an exponent in the denominator and as a decimal. $\dfrac{\square}{\square}$; $\square$

3. Complete the 1st 4 rows of the table showing the exponential and standard forms of power of 10.

4. What operation is performed when you move down the table?

_____

5. What happens to the exponent?

_____

6. Extend the table to include the next three entries.

Exponential Form	Standard Form
10^3	
$10^{\square}$	100
10^1	
10^0	

Essential Question

WHY is it helpful to write numbers in different ways?

Common Core State Standards

Content Standards
8.EE.1

MP Mathematical Practices
1, 3, 4, 7

Which MP **Mathematical Practices** did you use?
Shade the circle(s) that applies.

① Persevere with Problems
② Reason Abstractly
③ Construct an Argument
④ Model with Mathematics

⑤ Use Math Tools
⑥ Attend to Precision
⑦ Make Use of Structure
⑧ Use Repeated Reasoning

Zero and Negative Exponents

Words Any nonzero number to the zero power is 1. Any nonzero number to the negative *n* power is the multiplicative inverse of its *n*th power.

Examples Numbers

$5^0 = 1$

$7^{-3} = \frac{1}{7} \cdot \frac{1}{7} \cdot \frac{1}{7}$ or $\frac{1}{7^3}$

Algebra

$x^0 = 1, x \neq 0$

$x^{-n} = \frac{1}{x^n}, x \neq 0$

Work Zone

Negative Exponents
Remember that 6^{-3} is equal to $\frac{1}{6^3}$, not -216 or -18.

You can use exponents to represent very small numbers. Negative powers are the result of repeated division.

Examples

 Show your work.

Write each expression using a positive exponent.

1. 6^{-3}

$6^{-3} = \frac{1}{6^3}$ Definition of negative exponent

2. a^{-5}

$a^{-5} = \frac{1}{a^5}$ Definition of negative exponent

Got it? Do these problems to find out.

a. 7^{-2}

b. b^{-4}

c. 5^0

d. m^{-3}

Examples

Write each fraction as an expression using a negative exponent other than −1.

3. $\frac{1}{5^2}$

$\frac{1}{5^2} = 5^{-2}$ Definition of negative exponent

4. $\frac{1}{36}$

$\frac{1}{36} = \frac{1}{6^2}$ Definition of exponent

$= 6^{-2}$ Definition of negative exponent

Got it? Do these problems to find out.

e. $\frac{1}{8^3}$

f. $\frac{1}{4}$

g. $\frac{1}{c^5}$

h. $\frac{1}{27}$

a. _____

b. _____

c. _____

d. _____

e. _____

f. _____

g. _____

h. _____

Example

 Tutor

5. **STEM** One human hair is about 0.001 inch in diameter. Write the decimal as a power of 10.

$$0.001 = \frac{1}{1,000} \qquad \text{Write the decimal as a fraction.}$$

$$= \frac{1}{10^3} \qquad 1,000 = 10^3$$

$$= 10^{-3} \qquad \text{Definition of negative exponent}$$

A human hair is 10^{-3} inch thick.

Got it? Do this problem to find out.

i. **STEM** A water molecule is about 0.0000000001 meter long. Write the decimal as a power of 10.

i. _____

Multiply and Divide with Negative Exponents

The Product of Powers and the Quotient of Powers rules can be used to multiply and divide powers with negative exponents.

Examples

 Tutor

Simplify each expression.

6. $5^3 \cdot 5^{-5}$

$$5^3 \cdot 5^{-5} = 5^{3 + (-5)} \qquad \text{Product of Powers}$$

$$= 5^{-2} \qquad \text{Simplify.}$$

$$= \frac{1}{5^2} \text{ or } \frac{1}{25} \qquad \text{Write using positive exponents. Simplify.}$$

Show your work.

7. $\dfrac{w^{-1}}{w^{-4}}$

$$\frac{w^{-1}}{w^{-4}} = w^{-1 - (-4)} \qquad \text{Quotient of Powers}$$

$$= w^{(-1) + 4} \text{ or } w^3 \qquad \text{Subtract the exponents.}$$

Got it? Do these problems to find out.

j. $3^{-8} \cdot 3^2$

k. $\dfrac{11^2}{11^4}$

l. $n^9 \cdot n^{-4}$

m. $\dfrac{b^{-4}}{b^{-7}}$

j. _____

k. _____

l. _____

m. _____

Copyright © McGraw-Hill Education

STOP and Reflect

Explain below the difference between the expressions $(-4)^2$ and 4^{-2}.

Write each expression using a positive exponent. (Examples 1 and 2)

1. $2^{-4} =$ _____

2. $4^{-3} =$ _____

3. $a^{-4} =$ _____

4. $g^{-7} =$ _____

Write each fraction as an expression using a negative exponent other than −1.

(Examples 3 and 4)

5. $\dfrac{1}{3^4} =$ _____

6. $\dfrac{1}{m^5} =$ _____

7. $\dfrac{1}{16} =$ _____

8. $\dfrac{1}{49} =$ _____

9. An American green tree frog tadpole is about 0.00001 kilometer in length when it hatches. Write this decimal as a power of 10.

(Example 5) _____

Simplify. (Examples 6 and 7)

10. $3^{-3} \cdot 3^{-2} =$ _____

11. $r^{-7} \cdot r^3 =$ _____

12. $\dfrac{p^{-2}}{p^{-12}} =$ _____

13. **Building on the Essential Question** How are negative exponents and positive exponents related?

_____

_____

_____

_____

_____

_____

Rate Yourself!

How well do you understand understand negative exponents? Circle the image that applies.

Clear   Somewhat Clear   Not So Clear

For more help, go online to access a Personal Tutor.

Independent Practice

Go online for Step-by-Step Solutions

Write each expression using a positive exponent. (Examples 1 and 2)

1. $7^{-10} =$ _____

2. $(-5)^{-4} =$ _____

3. $g^{-7} =$ _____

4. $w^{-13} =$ _____

Write each fraction as an expression using a negative exponent other than --1.

(Examples 3 and 4)

5. $\dfrac{1}{12^4} =$ _____

6. $\dfrac{1}{(-5)^7} =$ _____

7. $\dfrac{1}{125} =$ _____

8. $\dfrac{1}{1,024} =$ _____

9. The table shows different metric measurements. Write each decimal as a power of 10. (Example 5) _____

Measurement	Value
Decimeter	0.1
Centimeter	0.01
Millimeter	0.001
Micrometer	0.000001

10. **STEM** An atom is a small unit of matter. A small atom measures about 0.0000000001 meter. Write the decimal as a power of 10.
(Example 5)

_____

Simplify. (Examples 6 and 7)

11. $2^{-3} \cdot 2^{-4} =$ _____

12. $s^{-5} \cdot s^{-2} =$ _____

13 $y^{-1} \cdot y^4 =$ _____

14. $(3a)(a^{-3}) =$ _____

15. $\dfrac{3^{-1}}{3^{-5}} =$ _____

16. $\dfrac{a^{-4}}{a^{-6}} =$ _____

17. $\dfrac{y^{-6}}{y^{-10}} =$ _____

18. $\dfrac{z^{-4}}{z^{-8}} =$ _____

19 STEM The mass of a molecule of penicillin is 10^{-18} kilogram and the mass of a molecule of insulin is 10^{-23} kilogram. How many times greater is the mass of a molecule of penicillin than the mass of a molecule of insulin?

_____

20. MP **Justify Conclusions** A common flea that is 2^{-4} inch long can jump about 2^3 inches high. About how many times its body size can a flea jump? Explain your reasoning.

_____

H.O.T. Problems Higher Order Thinking

21. MP **Identify Structure** Without evaluating, order 11^{-3}, 11^2, and 11^0 from least to greatest. Explain your reasoning.

_____

_____

22. MP **Identify Structure** Write an expression with a negative exponent that has a value between 0 and $\frac{1}{2}$.

_____

_____

23. MP **Persevere with Problems** Select several fractions between 0 and 1. Find the value of each fraction after it is raised to the -1 power. Explain the relationship between the -1 power and the original fraction.

_____

_____

_____

24. MP **Reason Abstractly** For each power, write an equivalent multiplication expression with two factors. The first factor should have a positive exponent and the second factor should have a negative exponent.

a. $10^4 = $ _____

b. $8^2 = $ _____

c. $x^7 = $ _____

Extra Practice

25. Write 3^{-5} using positive exponents. $\frac{1}{3^5}$

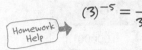

$(3)^{-5} = \frac{1}{3^5}$

26. Simplify $(4^{-4})(4^2)$. $\frac{1}{16}$

$(4^{-4})(4^2) = 4^{-4+2}$
$= 4^{-2}$
$= \frac{1}{4^2}$ or $\frac{1}{16}$

Write each expression using a positive exponent.

27. $6^{-8} =$ _____

28. $(-3)^{-5} =$ _____

29. $s^{-9} =$ _____

30. $t^{-11} =$ _____

Simplify.

31. $z^2 \cdot z^{-3} =$ _____

32. $n^{-1} \cdot n^3 =$ _____

33. $\frac{b^{-7}}{b^5} =$ _____

34. $\frac{x^4}{x^{-2}} =$ _____

35. $2^{-4} =$ _____

36. $(-5)^{-4} =$ _____

37. $(-10)^{-4} =$ _____

38. $(0.5)^{-4} =$ _____

MP Persevere with Problems Find the missing exponent.

39. $\frac{17^{\bullet}}{17^4} = 17^8$ _____

40. $\frac{k^6}{k^{\bullet}} = k^2$ _____

41. $\frac{p^{-1}}{p^{\bullet}} = p^{10}$ _____

42. The diameter of the average human cell is about 4^{-4} inch. Which of the following expressions are equivalent to this diameter? Select all that apply.

☐ $\frac{1}{4^4}$ in   ☐ $-\frac{1}{4^4}$ in.   ☐ $\frac{1}{256}$ in.   ☐ 0.00390625 in.

43. The table shows the values of different measurements in the metric system.

Select the correct answer to write each measurement as a power of 10.

Measurement	Value
micrometer	0.000001 m
millimeter	0.001 m
nanometer	0.000000001 m
picometer	0.000000000001 m

Measurement	Power of 10
micrometer	
millimeter	
nanometer	
picometer	

10^{-12}	10^{-5}
10^{-9}	10^{-3}
10^{-6}	10^{-2}

Common Core Spiral Review

Evaluate. 6.EE.1

44. $10^2 =$ _____

45. $10^3 =$ _____

46. $10^6 =$ _____

47. $10^5 =$ _____

Find each missing value. 6.NS.3

48. $0.003 \times$ _____ $= 3$

49. $0.079 \times$ _____ $= 7.9$

50. $0.00041 \times$ _____ $= 4.1$

51. $987 \div$ _____ $= 9.87$

52. $3,400 \div$ _____ $= 3.4$

53. $7,450 \div$ _____ $= 745$

Lesson 6
Scientific Notation

 Real-World Link

Electronics A single sided, single layer DVD has a storage capacity of 4.7 gigabytes. One gigabyte is equal to 10^9 bytes.

1. Write a multiplication expression that represents how many bytes can be stored on the DVD. _____

2. Complete the table below.

Expression	Product	Expression	Product
$4.7 \times 10^1 = 4.7 \times 10$	47	$4.7 \times 10^{-1} = 4.7 \times \frac{1}{10}$	0.47
$4.7 \times 10^2 = 4.7 \times 100$		$4.7 \times 10^{-2} = 4.7 \times \frac{1}{100}$	
$4.7 \times 10^3 = 4.7 \times 1,000$		$4.7 \times 10^{-3} = 4.7 \times \frac{1}{1000}$	
$4.7 \times 10^4 = 4.7 \times$ _____		$4.7 \times 10^{-4} = 4.7 \times$ _____	

3. If 4.7 is multiplied by a positive power of 10, what relationship exists between the decimal point's new position and the exponent?

_____

_____

4. When 4.7 is multiplied by a negative power of 10, how does the new position of the decimal point relate to the negative exponent? _____

_____

 Essential Question

WHY is it helpful to write numbers in different ways?

 Vocab **Vocabulary**

scientific notation

 Common Core State Standards

Content Standards
8.EE.4

 Mathematical Practices
1, 3, 4, 7

Which **Mathematical Practices** did you use?
Shade the circle(s) that applies.

① Persevere with Problems
② Reason Abstractly
③ Construct an Argument
④ Model with Mathematics

⑤ Use Math Tools
⑥ Attend to Precision
⑦ Make Use of Structure
⑧ Use Repeated Reasoning

Scientific Notation

Words **Scientific notation** is when a number is written as the product of a factor and an integer power of 10. When the number is positive the factor must be greater than or equal to 1 and less than 10.

Symbols $a \times 10^n$, where $1 \leq a < 10$ and n is an integer

Example $425,000,000 = 4.25 \times 10^8$

Use these rules to express a number in scientific notation.

- If the number is greater than or equal to 1, the power of ten is positive.
- If the number is between 0 and 1, the power of ten is negative.

Work Zone

Powers of Ten
Multiplying a factor by a positive power of 10 moves the decimal point right.
Multiplying a factor by a negative power of 10 moves the decimal point left.

a. _____

b. _____

c. _____

Examples

Tutor

Write each number in standard form.

1. 5.34×10^4

$5.34 \times 10^4 = 53,400.$

2. 3.27×10^{-3}

$3.27 \times 10^{-3} = 0.00327$

Got it? Do these problems to find out.

a. 7.42×10^5 b. 6.1×10^{-2} c. 3.714×10^2

Examples

Tutor

Write each number in scientific notation.

3. 3,725,000

$3,725,000 = 3.725 \times 1,000,000$ The decimal point moves 6 places.

$= 3.725 \times 10^6$ Since $3,725,000 > 1$, the exponent is positive.

- -

4. 0.000316

$0.000316 = 3.16 \times 0.0001$ The decimal point moves 4 places.

$= 3.16 \times 10^{-4}$ Since $0 < 0.000316 < 1$, the exponent is negative.

Show your work.

Got it? Do these problems to find out.

d. 14,140,000   e. 0.00876   f. 0.114

d. _____

e. _____

f. _____

 ## Example

5. Refer to the table at the right. Order the countries according to the amount of money visitors spent in the United States from greatest to least.

Dollars Spent by International Visitors in the U.S	
Country	Dollars Spent
Canada	1.03×10^7
India	1.83×10^6
Mexico	7.15×10^6
United Kingdom	1.06×10^7

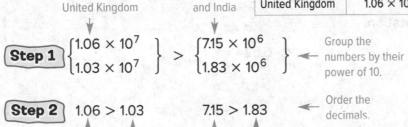

Canada and United Kingdom   Mexico and India

Step 1 $\begin{cases} 1.06 \times 10^7 \\ 1.03 \times 10^7 \end{cases} > \begin{cases} 7.15 \times 10^6 \\ 1.83 \times 10^6 \end{cases}$ ← Group the numbers by their power of 10.

Step 2 $1.06 > 1.03$ $7.15 > 1.83$ ← Order the decimals.

United Kingdom Canada   Mexico India

Got it? Do this problem to find out.

g. Some of the top U.S. cities visited by overseas travelers are shown in the table. Order the cities according to the number of visitors from least to greatest.

U.S. City	Number of Visitors
Boston	7.21×10^5
Las Vegas	1.3×10^6
Los Angeles	2.2×10^6
Metro D.C. area	9.01×10^5

g. _____

 ## Example

6. **STEM** If you could walk at a rate of 2 meters per second, it would take you 1.92×10^8 seconds to walk to the moon. Is it more appropriate to report this time as 1.92×10^8 seconds or 6.09 years? Explain your reasoning.

The measure 6.09 years is more appropriate. The number 1.92×10^8 seconds is very large so choosing a larger unit of measure is more meaningful.

Got it? **Do this problem to find out.**

h. _____

h. **STEM** In an ocean, the sea floor moved 475 kilometers over 65 million years. Is it more appropriate to report this rate as 7.31×10^{-5} kilometer per year or 7.31 centimeters per year? Explain your reasoning.

Guided Practice

Write each number in standard form. (Examples 1 and 2)

1. $9.931 \times 10^5 =$ _____

2. $6.02 \times 10^{-4} =$ _____

Write each number in scientific notation. (Examples 3 and 4)

3. $8,785,000,000 =$ _____

4. $0.524 =$ _____

5. The table lists the total value of music shipments for four years. List the years from least to greatest dollar amount.
(Example 5)

_____

_____

Year	Music Shipments ($)
1	1.22×10^{10}
2	1.12×10^{10}
3	7.15×10^6
4	1.06×10^7

6. **STEM** A plant cell has a diameter of 1.3×10^{-8} kilometer. Is it more appropriate to report the diameter of a plant cell as 1.3×10^{-8} kilometer or 1.3×10^{-2} millimeter? Explain your reasoning. (Example 6)

_____

_____

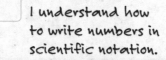

Rate Yourself!

☐ I understand how to write numbers in scientific notation.

 Great! You're ready to move on!

☐ I still have some questions about how to write numbers in scientific notation.

7. **ℚ Building on the Essential Question** How is scientific notation useful in the real world?

_____

_____

_____

_____

No Problem! Go online to access a Personal Tutor.

Independent Practice

Go online for Step-by-Step Solutions eHelp

Write each number in standard form. (Examples 1 and 2)

1. $3.16 \times 10^3 =$ _____

 Show your work.

2. $1.1 \times 10^{-4} =$ _____

3. $2.52 \times 10^{-5} =$ _____

Write each number in scientific notation. (Examples 3 and 4)

4. $43,000 =$ _____

5. $0.0072 =$ _____

6. $0.0000901 =$ _____

7. The areas of the world's oceans are listed in the table. Order the oceans according to their area from least to greatest. (Example 5)

_____

_____

World's Oceans	
Ocean	**Area (mi²)**
Atlantic	2.96×10^7
Arctic	5.43×10^6
Indian	2.65×10^7
Pacific	6×10^7
Southern	7.85×10^6

8. The space shuttle can travel about 8×10^5 centimeters per second. Is it more appropriate to report this rate as 8×10^5 centimeters per second or 8 kilometers per second? Explain. (Example 6)

_____

_____

_____

9. The inside diameter of a certain size of ring is 1.732×10^{-2} meter. Is it more appropriate to report the ring diameter as 1.732×10^{-2} meter or 17.32 millimeters? Explain. (Example 6)

_____

_____

_____

Fill in each $\bigcirc$ **with <, >, or = to make a true statement.**

10. $678,000 \bigcirc 6.78 \times 10^6$

11 $6.25 \times 10^3 \bigcirc 6.3 \times 10^3$

12. MP Model with Mathematics Refer to the graphic novel frame below for Exercises a–c.

a. Find Jacob's and Sarah's heights in nanometers.

_____

b. Write each height using scientific notation.

_____

c. Give an example of something that would be appropriately measured

by nanometers. _____

_____

H.O.T. Problems Higher Order Thinking

13. MP Justify Conclusions Determine whether 1.2×10^5 or 1.2×10^6 is

closer to one million. Explain. _____

_____

14. MP Persevere with Problems Compute and express each value in scientific notation.

a. $\dfrac{(130,000)(0.0057)}{0.0004} =$ _____

b. $\dfrac{(90,000)(0.0016)}{(200,000)(30,000)(0.00012)} =$ _____

15. MP Model with Mathematics Write two numbers in scientific notation with values between 100 and 1,000. Then write an inequality that shows the relationship between your two numbers.

_____

_____

Extra Practice

16. Write 7.113×10^7 in standard form.

71,130,000
_____

$7.113 \times 10^7 = 71130000.$ The decimal
⎵   point moves
7 places right.

Homework
Help

17. Write 0.00000707 in scientific notation.

7.07×10^{-6}
_____

$0.00000707 = 7.07 \times 0.000001$
$= 7.07 \times 10^{-6}$

The decimal point moves 6 places.
Since $0 < 0.00000707 < 1$,
the exponent is negative.

Write each number in standard form.

18. $2.08 \times 10^2 = $ _____

19. $7.8 \times 10^{-3} = $ _____

20. $8.73 \times 10^{-4} = $ _____

Write each number in scientific notation.

21. $6,700 = $ _____

22. $52,300,000 = $ _____

23. $0.037 = $ _____

24. **STEM** The table shows the mass in grams of one atom of each of
several elements. List the elements in order from the least mass to
greatest mass per atom.

_____

Element	Mass per Atom
Carbon	1.995×10^{-23} g
Gold	3.272×10^{-22} g
Hydrogen	1.674×10^{-24} g
Oxygen	2.658×10^{-23} g
Silver	1.792×10^{-22} g

MP Identify Structure Arrange each set of numbers in increasing order.

25. $216,000,000, 2.2 \times 10^3, 3.1 \times 10^7, 310,000$

_____

26. $4.56 \times 10^{-2}, 4.56 \times 10^3, 4.56 \times 10^2, 4.56 \times 10^{-3}$

_____

27. The thermosphere layer of the atmosphere is between 90,000 and 110,000 meters above sea level. Which of the following elevations are in the thermosphere? Select yes or no.

a. 9.8×10^{-4} ☐ Yes ☐ No

b. 1.04×10^{5} ☐ Yes ☐ No

c. 9.72×10^{4} ☐ Yes ☐ No

d. 1.45×10^{5} ☐ Yes ☐ No

28. The attendance for four Major League baseball teams for a recent year is shown below.

Sort the teams from least to greatest attendance.

Team	Attendance
Los Angeles Angels	3.06×10^{6}
Miami Marlins	22.2×10^{5}
Pittsburgh Pirates	20.9×10^{5}
St. Louis Cardinals	3.26×10^{6}

	Team	Attendance
Least		
Greatest		

Which team had the greatest attendance? ☐

Find each sum or difference. 6.NS.3

29. $9.7 + 0.532 =$ _____

30. $4.39 - 0.035 =$ _____

31. $679 - 1.4 =$ _____

Simplify. Express using exponents. 8.EE.1

32. $3a^4 \cdot 12a^2 =$ _____

33. $(5x)^2 \cdot 2x^5 =$ _____

34. $\dfrac{3^9}{3^2} =$ _____

Compute with Scientific Notation

Real-World Link

E-mail Every day, nearly 130 billion spam E-mails are sent worldwide! Use the steps below to find out how many are sent each year. The numbers are too large even for your calculator.

1. Express 130 billion in scientific notation.

2. Round 365 to the nearest hundred and express it in scientific notation.

3. Write a multiplication expression using the number in Exercises 1 and 2 to represent the total number of spam E-mails sent each year.

4. If you use the Commutative Property of Multiplication, you can rewrite the expression in Exercise 3 as $(1.3 \times 4)(10^{11} \times 10^{2})$. Evaluate this expression to find the number of spam E-mails sent in a year. Express the result in both scientific notation and standard form.

Essential Question

WHY is it helpful to write numbers in different ways?

CCSS **Common Core State Standards**

Content Standards
8.EE.3, 8.EE.4

MP Mathematical Practices
1, 3, 4

Which MP **Mathematical Practices** did you use?
Shade the circle(s) that applies.

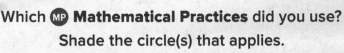

① Persevere with Problems

② Reason Abstractly

③ Construct an Argument

④ Model with Mathematics

⑤ Use Math Tools

⑥ Attend to Precision

⑦ Make Use of Structure

⑧ Use Repeated Reasoning

Multiplication and Division with Scientific Notation

You can use the Product of Powers and Quotient of Powers properties to multiply and divide numbers written in scientific notation.

Example
Tutor

1. Evaluate $(7.2 \times 10^3)(1.6 \times 10^4)$. Express the result in scientific notation.

$$(7.2 \times 10^3)(1.6 \times 10^4) = (7.2 \times 1.6)(10^3 \times 10^4)$$ Commutative and Associative Properties

$$= (11.52)(10^3 \times 10^4)$$ Multiply 7.2 by 1.6.

$$= 11.52 \times 10^{3\,+\,4}$$ Product of Powers

$$= 11.52 \times 10^7$$ Add the exponents.

$$= 1.152 \times 10^8$$ Write in scientific notation.

Got it? Do these problems to find out.

Show your work.

a. _____

b. _____

a. $(8.4 \times 10^2)(2.5 \times 10^6)$ **b.** $(2.63 \times 10^4)(1.2 \times 10^{-3})$

Real World

Example
Tutor

2. In 2010, the world population was about 6,860,000,000. The population of the United States was about 3×10^8. About how many times larger is the world population than the population of the United States?

Estimate the population of the world and write in scientific notation.

$$6{,}860{,}000{,}000 \approx 7{,}000{,}000{,}000 \text{ or } 7 \times 10^9$$

Find $\dfrac{7 \times 10^9}{3 \times 10^8}$.

$$\frac{7 \times 10^9}{3 \times 10^8} = \left(\frac{7}{3}\right)\left(\frac{10^9}{10^8}\right)$$ Associative Property

$$\approx 2.3 \times \left(\frac{10^9}{10^8}\right)$$ Divide 7 by 3. Round to the nearest tenth.

$$\approx 2.3 \times 10^{9\,-\,8}$$ Quotient of Powers

$$\approx 2.3 \times 10^1$$ Subtract the exponents.

So, the population of the world is about 23 times larger than the population of the United States.

Got it? Do this problem to find out.

c. The surface area of Lake Superior, the largest of the Great Lakes, is 8×10^4 square kilometers. The surface area of the smallest Great Lake, Ontario, is 18,160 square kilometers. About how many times as great is the area covered by Lake Superior than Lake Ontario?

c. _____

Addition and Subtraction with Scientific Notation

When adding or subtracting decimals in standard form, it is necessary to line up the place values. In scientific notation, the place value is represented by the exponent. Before adding or subtracting, both numbers must be expressed in the same form.

Examples

Evaluate each expression. Express the result in scientific notation.

3. $(6.89 \times 10^4) + (9.24 \times 10^5)$

$(6.89 \times 10^4) + (9.24 \times 10^5)$

$= (6.89 \times 10^4) + (92.4 \times 10^4)$ Write 9.24×10^5 as 92.4×10^4.

$= (6.89 + 92.4) \times 10^4$ Distributive Property

$= 99.29 \times 10^4$ Add 6.89 and 92.4.

$= 9.929 \times 10^5$ Rewrite in scientific notation.

4. $(7.83 \times 10^8) - 11,610,000$

$(7.83 \times 10^8) - (1.161 \times 10^7)$ Rewrite 11,610,000 in scientific notation.

$(7.83 \times 10^8) - (1.161 \times 10^7)$

$= (78.3 \times 10^7) - (1.161 \times 10^7)$ Write 7.83×10^8 as 78.3×10^7.

$= (78.3 - 1.161) \times 10^7$ Distributive Property

$= 77.139 \times 10^7$ Subtract 1.161 from 78.3.

$= 7.7139 \times 10^8$ Rewrite in scientific notation.

STOP and Reflect

Explain below how to estimate the sum of (4.215×10^{-2}) and (3.2×10^{-4}). Then find the estimate.

5. $593,000 + (7.89 \times 10^6)$

$593,000 + (7.89 \times 10^6)$

$= (5.93 \times 10^5) + (7.89 \times 10^6)$ Rewrite 593,000 in scientific notation.

$= (0.593 \times 10^6) + (7.89 \times 10^6)$ Write 5.93×10^5 as 0.593×10^6

$= (0.593 + 7.89) \times 10^6$ Distributive Property

$= 8.483 \times 10^6$ Add 0.593 and 7.89.

Show your work.

a. _____

e. _____

f. _____

Got it? Do these problems to find out.

d. $(8.41 \times 10^3) + (9.71 \times 10^4)$

e. $(1.263 \times 10^9) - (1.525 \times 10^7)$

f. $(6.3 \times 10^5) + 2,700,000$

Guided Practice

Check ✓

Evaluate each expression. Express the result in scientific notation. (Examples 1 and 2)

1. $(2.6 \times 10^5)(1.9 \times 10^2) =$ _____

2. $\dfrac{8.37 \times 10^8}{2.7 \times 10^3} =$ _____

Show your work.

3. In 2005, 8.1×10^{10} text messages were sent in the United States. In 2010, the number of annual text messages had risen to 1,810,000,000,000. About how many times as great was the number of text messages in 2010 than 2005?
(Example 2)

_____

Evaluate each expression. Express the result in scientific notation. (Examples 3–5)

4. $(8.9 \times 10^9) + (4.2 \times 10^6) =$ _____

5. $(9.64 \times 10^8) - (5.29 \times 10^6) =$ _____

6. $(1.35 \times 10^6) - (117,000) =$ _____

7. $5,400 + (6.8 \times 10^5) =$ _____

8. ⓔ **Building on the Essential Question** How does scientific notation make it easier to perform computations with very large or very small numbers? _____

_____

_____

Rate Yourself!

Are you ready to move on?
Shade the section that applies.

YES   ?   NO

For more help, go online to access a Personal Tutor.

Tutor 💬

Independent Practice

Go online for Step-by-Step Solutions
eHelp

Evaluate each expression. Express the result in scientific notation. (Examples 1 and 2)

1. $(3.9 \times 10^2)(2.3 \times 10^6) =$ _____

Show your work.

2. $(4.18 \times 10^{-4})(9 \times 10^{-4}) =$ _____

3. $(9.75 \times 10^3)(8.4 \times 10^{-6}) =$ _____

4. $\dfrac{9.45 \times 10^{10}}{1.5 \times 10^6} =$ _____

5. $\dfrac{1.14 \times 10^6}{4.8 \times 10^{-6}} =$ _____

6. $\dfrac{9 \times 10^{-11}}{2.4 \times 10^8} =$ _____

7. **STEM** Neurons are cells in the nervous system that process and transmit information. An average neuron is about 5×10^{-6} meter in diameter. A standard table tennis ball is 0.04 meter in diameter. About how many times as great is the diameter of a ball than a neuron? (Example 2)

_____

_____

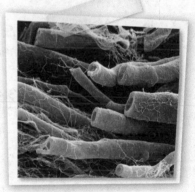

Evaluate each expression. Express the result in scientific notation.

(Examples 3–5)

8. $(9.5 \times 10^{11}) + (6.3 \times 10^9) =$ _____

9. $(1.03 \times 10^9) - (4.7 \times 10^7) =$ _____

10. $(1.357 \times 10^9) + 590,000 =$ _____

11. $87,100 - (6.34 \times 10^1) =$ _____

12. **Persevere with Problems** Central Park in New York City is rectangular in shape and measures approximately 1.37×10^4 feet by 2.64×10^2 feet. If one acre is equal to 4.356×10^4 square feet, how many acres does Central Park cover? Round to the nearest hundredth.

_____

H.O.T. Problems Higher Order Thinking

13. **MP Find the Error** Enrique is finding $\dfrac{6.63 \times 10^{-6}}{5.1 \times 10^{-2}}$. Circle his mistake and correct it.

_____

_____

_____

_____

$$\frac{6.63 \times 10^{-6}}{5.1 \times 10^{-2}} = \left(\frac{6.63}{5.1}\right)\left(\frac{10^{-6}}{10^{-2}}\right)$$
$$= 1.3 \times 10^{-6-2}$$
$$= 1.3 \times 10^{-8}$$

14. **MP Which One Doesn't Belong?** Identify the expression that does not belong with the other three. Explain your reasoning.

| 14.28×10^9 | $(3.4 \times 10^6)(4.2 \times 10^3)$ | 1.4×10^9 | $(3.4)(4.2) \times 10^{(6+3)}$ |

_____

_____

15. **MP Persevere with Problems** A *googol* is the number 1 followed by 100 zeros.

   a. What is one googol written in scientific notation? _____

   b. How many times greater is a googol of meters than a nanometer? _____

   c. There are about 2.5×10^{10} red blood cells in the average adult. About how many adults would it take to have a total of 1 googol red blood cells? _____

16. **MP Model with Mathematics** Write an addition expression and a subtraction expression, each with a value of 2.4×10^{-3}.

_____

_____

Extra Practice

Evaluate each expression. Express the result in scientific notation.

17. $(3.7 \times 10^{-2})(1.2 \times 10^3) = \underline{4.44 \times 10^1}$

$(3.7 \times 10^{-2})(1.2 \times 10^3) = (3.7 \times 1.2) \times$
$(10^{-2} \times 10^3)$

Homework Help →

$= 4.44 \times 10^{-2+3}$
$= 4.44 \times 10^1$

18. $\dfrac{4.64 \times 10^{-4}}{2.9 \times 10^{-6}} = \underline{1.6 \times 10^2}$

$\dfrac{4.64 \times 10^{-4}}{2.9 \times 10^{-6}} = \dfrac{4.64}{2.9} \times \dfrac{10^{-4}}{10^{-6}}$

$= 1.6 \times 10^{-4-(-6)}$
$= 1.6 \times 10^2$

19. $\dfrac{3.24 \times 10^{-4}}{8.1 \times 10^{-7}} = $ _____

20. $(7.3 \times 10^5) + 2{,}400{,}000 = $ _____

21. $(8.64 \times 10^6) + (1.334 \times 10^{10}) = $

_____

22. $(1.21 \times 10^5) - 9{,}500 = $

_____

23. **MP Persevere with Problems** A circular swimming pool holds 1.22×10^6 cubic inches of water. It is being filled at a rate of 1.5×10^3 cubic inches per minute. How many hours will it take to fill the swimming pool? _____

_____

24. **Financial Literacy** In 2010, the national debt of the United States was about 14 trillion dollars. In 2003 it was about 7×10^{12} dollars. About how many times larger was the national debt in 2010 than in 2003? _____

_____

25. There are approximately 45 hundred species of mammals on Earth and 2.8×10^4 species of fish. Fill in each box to make a true statement.

There are more species of [_____] than species of [_____]

on Earth. The difference in the number of species is [_____].

26. *Population density* is a measure of how many people are living in a region. To calculate population density, divide the population of a region by the area in square miles. The table shows the approximate populations and areas of different countries.

Country	Population	Area (mi²)
China	1.332×10^9	3.7×10^6
Poland	3.84×10^7	1.2×10^5
Sweden	9.6×10^6	1.6×10^5
United States	3.15×10^8	3.5×10^6

Sort the countries from least to greatest population density.

	Country	Population Density Density (people per mi²)
Least		
Greatest		

Which country has the greatest population density? [_____]

Common Core Spiral Review

27. A cube measures 6.6 inches on each side. **6.G.2**

a. Find the area of one face of the cube. _____

b. Find the volume of the cube. _____

28. Complete the table shown. **6.EE.1**

x	x^2	x^3	x	x^2	x^3
1			7		
2			8		
3			9		
4			10		
5			11		
6			12		

Inquiry Lab

Graphing Technology: Scientific Notation Using Technology

 WHAT are the similarities and differences between a number written in scientific notation and the calculator notation of the number shown on a screen?

 Content Standards 8.EE.3, 8.EE.4

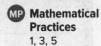 **Mathematical Practices** 1, 3, 5

The table shows the mass of some planets in our solar system. What is the mass of Earth written in scientific notation?

Planet	Mass (kg)
Earth	5,973,700,000,000,000,000,000,000
Mars	641,850,000,000,000,000,000,000
Saturn	568,510,000,000,000,000,000,000,000

What do you know? _____

What do you need to find? _____

Hands-On Activity 1

You will use a graphing calculator to explore how scientific notation is displayed using technology.

Step 1 Press CLEAR to clear the home screen.

Step 2 Enter the value in standard form for Earth's mass. Press ENTER.

Copy your calculator screen on the blank screen shown.

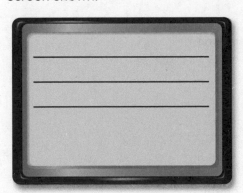

Step 3 Write the value for Earth's mass using scientific notation.

_____

Investigate

MP Use Math Tools Work with a partner. Repeat Steps 1 and 2 of the activity on the previous page for each of the following.

1. mass of Mars

2. mass of Saturn

Show your work.

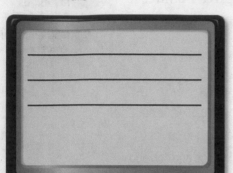

3. What does the E symbol represent on the calculator screen?

What does the value after the E symbol represent? _____

4. Based on your answer for Exercise 1, what is the mass of Mars in scientific notation? _____

5. Based on your answer for Exercise 2, what is the mass of Saturn in scientific notation? _____

Analyze and Reflect

Collaborate

Work with a partner to complete the table.

	Calculator Notation	Scientific Notation	Standard Form
6.	3.1E7		
7.		6.39×10^{10}	
8.			0.02357
9.	1.7E-11		

10. **MP Reason Inductively** The Moon has a mass of about 73,600,000,000,000,000,000,000 kilograms. Without entering the value in your calculator, predict how the mass of the Moon will be displayed on the calculator screen. _____

Hands-On Activity 2

A human blood cell is about 1×10^{-6} meter in diameter. The Moon is about 3.476×10^6 meters in diameter. How many times greater is the diameter of the Moon than the diameter of a blood cell?

Step 1 Press CLEAR to clear the home screen.

Step 2 Perform the following keystrokes:

3.476 2nd [EE] **6** ÷ **1** 2nd [EE] **−6** ENTER

Copy your calculator screen on the blank screen shown.

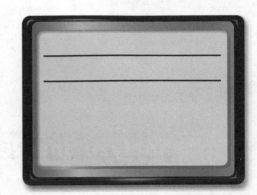

Step 3 Write the value in standard form.

_____

So, the diameter of the Moon is _____ times greater than the diameter of a human blood cell.

Hands-On Activity 3

When in "Normal" mode, a calculator will show answers in scientific notation only if they are very large numbers or very small numbers. You can set your calculator to show scientific notation for all numbers by using the "Sci" mode.

Step 1 Press CLEAR to clear the home screen. Put your calculator in scientific mode by pressing MODE ▶ ENTER. Then press CLEAR to return to the home screen.

Step 2 Complete the table by entering the numbers in the first column into your calculator.

Enter	Calculator Notation	Standard Form
$14 \div 100$		
$60 - 950$		
$360 \cdot 15$		
$1 + 1$		

11. **MP Use Math Tools** Work with a partner. Write down the keystrokes and fill in the calculator screen to find $(6.2 \times 10^5)(2.3 \times 10^7)$ using a calculator in "Sci" mode. Write your final answer in standard form.

 Show your work.

Keystrokes: _____

_____

Answer in standard form: _____

 Analyze and Reflect

12. **MP Use Math Tools** A *micrometer* is 0.000001 meter. Use your calculator to determine how many micrometers are in each of the following. Write your answer in both calculator and scientific notation.

	Calculator Notation	Scientific Notation
5,000 meters		
4.08E14 meters		
2.9E-10 meter		

Create

On Your Own

13. **MP Use Math Tools** Write a subtraction expression involving two numbers written in scientific notation. Then write the keystrokes and fill in the calculator screen to find the answer using a calculator in "Sci" mode. Write your final answer in standard form.

Expression: _____

Keystrokes: _____

_____

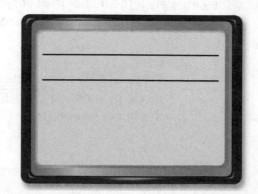

Answer in standard form: _____

14. **Inquiry** WHAT are the similarities and differences between a number written in scientific notation and the calculator notation of the number shown on a screen? _____

_____

_____

Roots

Vocabulary Start-Up

A **square root** of a number is one of its two equal factors. Numbers such as 1, 4, 9, 16, and 25 are called **perfect squares** because they are squares of integers.

Complete the graphic organizer.

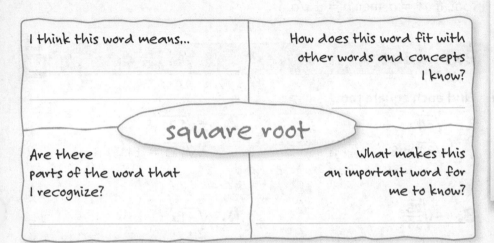

I think this word means...

_____

_____

How does this word fit with other words and concepts I know?

Are there parts of the word that I recognize?

_____

square root

What makes this an important word for me to know?

What is the relationship between squaring a number and finding the square root? _____

 Real-World Link Watch ▶

The square base of the Great Pyramid of Giza covers almost 562,500 square feet. How could you determine the length of each side of the base?

_____

Which **MP Mathematical Practices** did you use?
Shade the circle(s) that applies.

① Persevere with Problems

② Reason Abstractly

③ Construct an Argument

④ Model with Mathematics

⑤ Use Math Tools

⑥ Attend to Precision

⑦ Make Use of Structure

⑧ Use Repeated Reasoning

Essential Question

WHY is it helpful to write numbers in different ways?

Vocabulary

square root
perfect square
radical sign
cube root
perfect cube

Common Core State Standards

Content Standards
8.EE.2

MP Mathematical Practices
1, 3, 4

Square Root

Watch

Words A square root of a number is one of its two equal factors.

Symbols If $x^2 = y$, then x is a square root of y.

Example $5^2 = 25$ so 5 is a square root of 25.

Work Zone

Every positive number has *both* a positive and negative square root. In most real-world situations, only the positive or *principal* square root is considered. A **radical sign**, $\sqrt{}$, is used to indicate the principal square root. If $n^2 = a$, then $n = \pm\sqrt{a}$.

Examples

Tutor

Find each square root.

1. $\sqrt{64}$

$\sqrt{64} = 8$ Find the positive square root of 64; $8^2 = 64$.

2. $\pm\sqrt{1.21}$

$\pm\sqrt{1.21} = \pm1.1$ Find both square roots of 1.21; $1.1^2 = 1.21$.

Show your work.

3. $-\sqrt{\dfrac{25}{36}}$

$-\sqrt{\dfrac{25}{36}} = -\dfrac{5}{6}$ Find the negative square root of $\dfrac{25}{36}$; $\left(\dfrac{5}{6}\right)^2 = \dfrac{25}{36}$.

4. $\sqrt{-16}$

There is no real square root because no number times itself is equal to -16.

a. _____

b. _____

c. _____

Got it? Do these problems to find out.

a. $\sqrt{\dfrac{9}{16}}$ b. $\pm\sqrt{0.81}$ c. $-\sqrt{49}$ d. $\sqrt{-100}$

d. _____

Example

Tutor

5. Solve $t^2 = 169$. Check your solution(s).

$t^2 = 169$ Write the equation.

$t = \pm\sqrt{169}$ Definition of square root

$t = 13$ and -13 Check $13 \cdot 13 = 169$ and $(-13)(-13) = 169$ ✓

e. _____

f. _____

Got it? Do these problems to find out.

g. _____

e. $289 = a^2$ f. $m^2 = 0.09$ g. $y^2 = \dfrac{4}{25}$

Cube Roots

Words   A **cube root** of a number is one of its three equal factors.

Symbols   If $x^3 = y$, then x is the cube root of y.

Numbers such as 8, 27, and 64 are **perfect cubes** because they are the cubes of integers.

$8 = 2 \cdot 2 \cdot 2$ or 2^3 $27 = 3 \cdot 3 \cdot 3$ or 3^3 $64 = 4 \cdot 4 \cdot 4$ or 4^3

The symbol $\sqrt[3]{}$ is used to indicate a cube root of a number.

If $n^3 = a$, then $n = \sqrt[3]{a}$. You can use this relationship to solve equations that involve cubes.

Examples

Find each cube root.

6. $\sqrt[3]{125}$

$\sqrt[3]{125} = 5$ $5^3 = 5 \cdot 5 \cdot 5$ or 125

7. $\sqrt[3]{-27}$

$\sqrt[3]{-27} = -3$ $(-3)^3 = (-3) \cdot (-3) \cdot (-3)$ or -27

Got it? Do these problems to find out.

h. $\sqrt[3]{729}$ **i.** $\sqrt[3]{-64}$ **j.** $\sqrt[3]{1,000}$

Cube Roots

While $\sqrt{-16}$ is not a real number, $\sqrt[3]{-27}$ is a real number. $-3 \cdot -3 \cdot -3 = -27$

Show your work.

h. _____

i. _____

j. _____

Example

8. Dylan has a planter in the shape of a cube that holds 8 cubic feet of potting soil. Solve the equation $8 = s^3$ to find the side length s of the container.

$8 = s^3$ Write the equation.

$\sqrt[3]{8} = s$ Take the cube root of each side.

$2 = s$ Definition of cube root

So, each side of the container is 2 feet.

Check $(2)^3 = 8$ ✓

k. _____

k. An aquarium in the shape of a cube that will hold 25 gallons of water has a volume of 3.375 cubic feet. Solve $s^3 = 3.375$ to find the length of one side of the aquarium.

Guided Practice

Check ✓

Find each square root. (Examples 1–4)

Show your work.

1. $-\sqrt{1.69} =$ _____

2. $\pm\sqrt{\dfrac{49}{144}} =$ _____

3. $\sqrt{-1.44} =$ _____

Solve each equation. Check your solution(s). (Example 5)

4. $p^2 = 36$ _____

5. $t^2 = \dfrac{1}{9}$ _____

6. $6.25 = r^2$ _____

Find each cube root. (Examples 6 and 7)

7. $\sqrt[3]{216} =$ _____

8. $\sqrt[3]{-125} =$ _____

9. $\sqrt[3]{-8} =$ _____

10. A cube-shaped packing box can hold 729 cubic inches of packing material. Solve $729 = s^3$ to find the length of one side of the box. (Example 8) _____

11.  **Building on the Essential Question** Why would I need to use square roots and cube roots?

_____

_____

_____

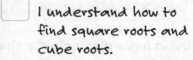
Rate Yourself!

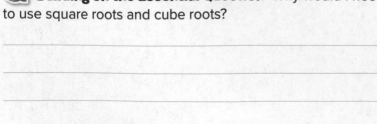
☐ I understand how to find square roots and cube roots.

▷▷ Great! You're ready to move on!

☐ I still have some questions about finding square roots and cube roots.

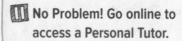
No Problem! Go online to access a Personal Tutor. Tutor

Independent Practice

Go online for Step-by-Step Solutions

Find each square root. (Examples 1–4)

1. $\sqrt{16} =$ _____

2. $-\sqrt{484} =$ _____

3. $\sqrt{-36} =$ _____

4. $\pm\sqrt{\dfrac{9}{49}} =$ _____

5 $-\sqrt{2.56} =$ _____

6. $\sqrt{-0.25} =$ _____

Solve each equation. Check your solution(s). (Example 5)

7. $v^2 = 81$ _____

8. $w^2 = \dfrac{36}{100}$ _____

9. $0.0169 = c^2$ _____

Find each cube root. (Examples 6 and 7)

10. $\sqrt[3]{1,728} =$ _____

11. $\sqrt[3]{-0.125} =$ _____

12. $\sqrt[3]{\dfrac{27}{125}} =$ _____

13 A group of 169 students needs to be seated in a square formation for a yearbook photo. Solve the equation $169 = s^2$ to find how many students

should be in each row. (Example 8) _____

14. Chloe wants to build a storage container in the shape of a cube to hold 15.625 cubic meters of hay for her horse. Solve the equation $15.625 = s^3$ to find the length of one side of the container. (Example 8)

_____

MP Persevere with Problems Given the area of each square, find the perimeter.

15.

Area = 121 square inches

_____

16.

Area = 25 square feet

_____

17.

Area = 36 square meters

_____

 H.O.T. Problems Higher Order Thinking

MP Persevere with Problems Find each value.

18. $\left(\sqrt{36}\right)^2 =$ _____

19. $\left(\sqrt{\frac{25}{81}}\right)^2 =$ _____

20. $\left(\sqrt{199}\right)^2 =$ _____

21. $\left(\sqrt{x}\right)^2 =$ _____

22. MP Reason Abstractly Based on your solutions to Exercises 18–21, write a rule that could be used to simplify the square of any square root of a number.

_____

23. MP Reason Inductively Explain why $\sqrt{-4}$ is not a real number, but $\sqrt[3]{-8}$ is.

_____

_____

24. MP Reason Inductively Describe the difference between an exact value and an approximation when finding square roots of numbers that are not perfect squares. Give an example of each.

_____

_____

_____

Extra Practice

Find each square root.

25. $-\sqrt{81} =$ _−9_

Homework Help → $9 \cdot 9 = 81$

So, $-\sqrt{81} = -9$.

26. $-\sqrt{\dfrac{64}{225}} =$ _____

27. $-\sqrt{\dfrac{16}{25}} =$ _____

28. $\pm\sqrt{1.44} =$ _____

Find each cube root.

29. $\sqrt[3]{-216} =$ _____

30. $\sqrt[3]{-512} =$ _____

31. $\sqrt[3]{-1,000} =$ _____

32. $\sqrt[3]{-343} =$ _____

Solve each equation. Check your solution(s).

33. $b^2 = 100$

34. $\dfrac{9}{64} = c^2$

35. $a^2 = 1.21$

36. $\dfrac{1}{8} = z^3$

37. $1.331 = c^3$

38. $m^3 = 8,000$

39. $\sqrt{x} = 5$

40. $\sqrt{y} = 20$

41. $\sqrt{z} = 10.5$

42. **MP** **Persevere with Problems** A concert crew needs to set up some chairs on the floor level. The chairs are to be placed in a square pattern consisting of four square sections. If one of the square sections holds 900 chairs, how many chairs will there be along each length of the larger square? _____

43. Mr. Freeman has a square cornfield. Which of the following could be the area of the cornfield if the sides are measured in whole numbers? Select all that apply.

☐ 164,000 ft²   ☐ 156,816 ft²   ☐ 174,724 ft²   ☐ 215,908 ft²

44. The area of each square in the figures below is 81 square units. Select the perimeter of each figure.

88 units	99 units
90 units	108 units
94 units	117 units

Do any of the figures have the same perimeter? If so, explain why.

Common Core Spiral Review

Evaluate each expression. 8.EE.2

45. $13^3 = $ _____

46. $25^2 = $ _____

47. $15^3 = $ _____

48. $34^2 = $ _____

49. $5 \cdot \sqrt{121} = $ _____

50. $-6 \cdot \sqrt{36} = $ _____

51. $10 \cdot \sqrt[3]{8} = $ _____

52. $-4 \cdot \sqrt{144} = $ _____

Express the volume of each cube as a monomial. 8.EE.1

53. 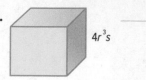 _____ $4r^3s$

54. 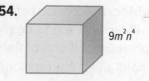 $9m^2n^4$

Inquiry Lab

Roots of Non-Perfect Squares

 Inquiry HOW can you estimate the square root of a non-perfect square number?

CCSS Content Standards 8.NS.2, 8.EE.2

MP Mathematical Practices 1, 3, 4, 5

Mindi is making a quilting piece from a square pattern as shown. Each of the dotted lines is 6 inches. What is the approximate length of one side of the square?

What do you know? _____

What do you need to find? _____

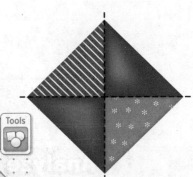

Hands-On Activity

Tools

Step 1 The outline of the square on dot paper is shown. Draw dotted lines connecting opposite vertices.

When you draw the lines, four triangles that are the same shape and size are formed. What are the dimensions of the triangles?

base = ☐ units   height = ☐ units

The area of one triangle is ☐ square units.

The area of the square is ☐ square units.

Step 2 Copy and cut out the square in Step 1 on another sheet of paper.

Step 3 Place one side of your square on the number line. Between what two consecutive whole numbers is $\sqrt{18}$, the side length of the square,

located? _____

0 1 2 3 4 5 6 7 8 9 10

The side of the square is closer to which one of the two whole

numbers? _____   Estimate $\sqrt{18}$. _____

So, one side of the square is about ☐ units long.

Investigate

MP Use Math Tools Work with a partner. Determine the two consecutive whole numbers the side length of each square is located between using the method shown in the Investigation.

1. _____

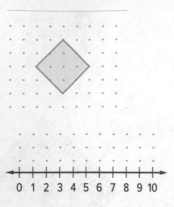

0 1 2 3 4 5 6 7 8 9 10

2. _____

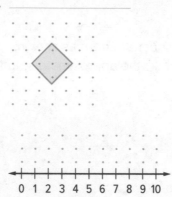

0 1 2 3 4 5 6 7 8 9 10

3. _____

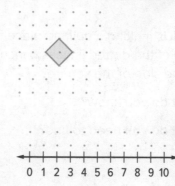

0 1 2 3 4 5 6 7 8 9 10

Analyze and Reflect

MP Use Math Tools Estimate the side length of each square in Exercises 1–3. Verify your estimate by using a calculator.

4. Exercise 1

Estimate _____

Calculator _____

5. Exercise 2

Estimate _____

Calculator _____

6. Exercise 3

Estimate _____

Calculator _____

7. MP Reason Inductively How does using a square model help you find

the square root of a non-perfect square? _____

_____

Create

8. MP Model with Mathematics Write a real-world problem that involves estimating a square root. Use the method shown in the activity to solve your problem.

_____

_____

9. **Inquiry** HOW can you estimate the square root of a non-perfect square?

_____

_____

Estimate Roots

 Real-World Link

Watch

Gravity Legend states that while sitting in his garden one day, Sir Isaac Newton was struck on the head by an apple. Suppose the apple was 25 feet above his head. Use the following steps to find how long it took the apple to fall.

1. What is the square root of 25?

2. The formula $t = \dfrac{\sqrt{h}}{4}$ can be used to find the time t in seconds it will take an object to fall from a certain height h in feet. How long did it take the apple to fall?

_____

3. Suppose another apple was 13 feet above the ground. Use the formula to write an equation representing the time it would have taken for the apple to hit the ground.

_____

4. Can you write $\dfrac{\sqrt{13}}{4}$ without a radical sign? Explain.

_____
_____
_____
_____

 Essential Question

WHY is it helpful to write numbers in different ways?

 Common Core State Standards

Content Standards
8.NS.2, 8.EE.2

MP Mathematical Practices
1, 3, 4

Which **MP Mathematical Practices** did you use?
Shade the circle(s) that applies.

① Persevere with Problems
② Reason Abstractly
③ Construct an Argument
④ Model with Mathematics

⑤ Use Math Tools
⑥ Attend to Precision
⑦ Make Use of Structure
⑧ Use Repeated Reasoning

Estimate Square and Cube Roots

You know that $\sqrt{8}$ is not a whole number because 8 is not a perfect square. The number line below shows that $\sqrt{8}$ is between 2 and 3. Since 8 is closer to 9 than 4, the best whole number estimate for $\sqrt{8}$ is 3.

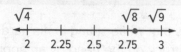

Examples

1. Estimate $\sqrt{83}$ to the nearest integer.

- The largest perfect square less than 83 is 81. $\sqrt{81} = 9$
- The smallest perfect square greater than 83 is 100. $\sqrt{100} = 10$

Plot each square root on a number line. Then estimate $\sqrt{83}$.

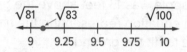

$$81 < 83 < 100 \qquad \text{Write an inequality.}$$
$$9^2 < 83 < 10^2 \qquad 81 = 9^2 \text{ and } 100 = 10^2$$
$$\sqrt{9^2} < \sqrt{83} < \sqrt{10^2} \qquad \text{Find the square root of each number.}$$
$$9 < \sqrt{83} < 10 \qquad \text{Simplify.}$$

So, $\sqrt{83}$ is between 9 and 10. Since $\sqrt{83}$ is closer to $\sqrt{81}$ than $\sqrt{100}$, the best integer estimate for $\sqrt{83}$ is 9.

Inequalities

$81 < 83 < 100$ is read 81 is less than 83 which is less than 100, or 83 is between 81 and 100.

2. Estimate $\sqrt[3]{320}$ to the nearest integer.

- The largest perfect cube less than 320 is 216. $\sqrt[3]{216} = 6$
- The smallest perfect cube greater than 320 is 343. $\sqrt[3]{343} = 7$

$$216 < 320 < 343 \qquad \text{Write an inequality.}$$
$$6^3 < 320 < 7^3 \qquad 216 = 6^3 \text{ and } 343 = 7^3$$
$$\sqrt[3]{6^3} < \sqrt[3]{320} < \sqrt[3]{7^3} \qquad \text{Find the cube root of each number.}$$
$$6 < \sqrt[3]{320} < 7 \qquad \text{Simplify.}$$

So, $\sqrt[3]{320}$ is between 6 and 7. Since 320 is closer to 343 than 216, the best integer estimate for $\sqrt[3]{320}$ is 7.

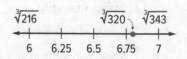

Got it? Do these problems to find out.

a. $\sqrt{35}$ b. $\sqrt{170}$ c. $\sqrt{44.8}$

d. $\sqrt[3]{62}$ e. $\sqrt[3]{25}$ f. $\sqrt[3]{129.6}$

Show your work.

a. _____

b. _____

c. _____

d. _____

e. _____

f. _____

 Real World

Example

 Tutor

3. Wyatt wants to fence in a square portion of the yard to make a play area for his new puppy. The area covered is 2 square meters. How much fencing should Wyatt buy?

$\sqrt{2}$ m
2 m² $\sqrt{2}$ m

Wyatt will need $4 \cdot \sqrt{2}$ meters of fencing. The square root of 2 is between 1 and 2 so $4 \cdot \sqrt{2}$ is between 4 and 8. Is this the best approximation? You can truncate the decimal expansion of $\sqrt{2}$ to find better approximations.

Estimate $\sqrt{2}$ by truncating, or dropping, the digits after the first decimal place, then after the second decimal place, and so on until an appropriate approximation is reached.

$\sqrt{2} \approx 1.414213562$ Use a calculator.

$\sqrt{2} \approx 1.4\cancel{14213562}$ Truncate, or drop, the digits after the first decimal place. $\sqrt{2}$ is between 1.4 and 1.5.

$5.6 < 4\sqrt{2} < 6.0$ $4 \cdot 1.4 = 5.6$ and $4 \cdot 1.5 = 6.0$

To find a closer approximation, expand $\sqrt{2}$ then truncate the decimal expansion after the first two decimal places.

$\sqrt{2} \approx 1.41\cancel{4213562}$ $\sqrt{2}$ is between 1.41 and 1.42.

$5.64 < 4\sqrt{2} < 5.68$ $4 \cdot 1.41 = 5.64$ and $4 \cdot 1.42 = 5.68$

The approximations indicate that Wyatt should buy 6 meters of fencing.

 **STOP** and Reflect

What is the difference between an exact value and an approximate value when finding square roots of numbers that are not perfect squares? Explain below.

Got it? Do this problem to find out.

g. Kelly needs to put trim around a circular tablecloth with a diameter of 36 inches. Use the equation $C = \pi d$ to find three sets of approximations for the amount of trim she will need. Truncate the value of π to the ones, tenths, and hundredths place. Then determine how much trim she should buy.

g. _____

4. The *golden rectangle* is found frequently in the nautilus shell. The length of the longer side divided by the length of the shorter side is equal to $\frac{1 + \sqrt{5}}{2}$. Estimate this value.

First estimate the value of $\sqrt{5}$.

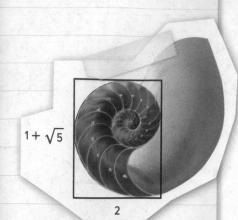

$1 + \sqrt{5}$

2

$$4 < 5 < 9 \qquad \text{4 and 9 are the closest perfect squares.}$$
$$2^2 < 5 < 3^2 \qquad 4 = 2^2 \text{ and } 9 = 3^2$$
$$\sqrt{2^2} < \sqrt{5} < \sqrt{3^2} \qquad \text{Find the square root of each number.}$$
$$2 < \sqrt{5} < 3 \qquad \text{Simplify.}$$

Since 5 is closer to 4 than 9, the best integer estimate for $\sqrt{5}$ is 2. Use this value to evaluate the expression.

$$\frac{1 + \sqrt{5}}{2} \approx \frac{1 + 2}{2} \text{ or } 1.5$$

Guided Practice

Check ✓

Estimate to the nearest integer. (Examples 1 and 2)

Show your work.

1. $\sqrt{28} \approx$ _____

2. $\sqrt{135} \approx$ _____

3. $\sqrt{38.7} \approx$ _____

4. $\sqrt[3]{51} \approx$ _____

5. $\sqrt[3]{200} \approx$ _____

6. $\sqrt[3]{95} \approx$ _____

7. **STEM** Tobias dropped a tennis ball from a height of 60 meters. The time in seconds it takes for the ball to fall 60 feet is $0.25(\sqrt{60})$. Find three sets of approximations for the amount of time it will take. Then determine how long it will take for the ball to hit the ground. (Example 3)

_____

_____

8. The number of swings back and forth of a pendulum of length L in inches each minute is $\frac{375}{\sqrt{L}}$. About how many swings will a 40-inch pendulum make each minute? (Example 4)

_____

9. **Building on the Essential Question** How can I estimate the square root of a non-perfect square?

_____

_____

Rate Yourself!

How confident are you about finding the square root of a non-perfect square? Mark an X in the section that applies.

I'm on target.

I need help.

For more help, go online to access a Personal Tutor.

Tutor

Independent Practice

Go online for Step-by-Step Solutions

Estimate to the nearest integer. (Examples 1 and 2)

1. $\sqrt{23} \approx$ _____

Show your work.

2. $\sqrt{197} \approx$ _____

3. $\sqrt{15.6} \approx$ _____

4. $\sqrt{85.1} \approx$ _____

5. $\sqrt[3]{22} \approx$ _____

6. $\sqrt[3]{34} \approx$ _____

7 $\sqrt[3]{989} \approx$ _____

8. $\sqrt[3]{250} \approx$ _____

9. The area of Kaitlyn's square garden is 345 square feet. One side of the garden is next to a shed. She wants to put a fence around the other three sides of the garden. Find three sets of approximations for the amount of fence it will take. Then determine how much fence she should buy.

(Example 3) _____

_____

10. In Little League, the bases are squares with sides of 14 inches. The expression $\sqrt{(s^2 + s^2)}$ represents the distance *diagonally across* a square of side length *s*. Estimate the diagonal distance across a base to

the nearest inch. (Example 4) _____

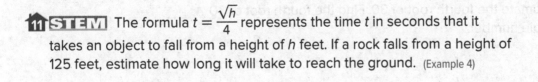
11 **STEM** The formula $t = \dfrac{\sqrt{h}}{4}$ represents the time *t* in seconds that it takes an object to fall from a height of *h* feet. If a rock falls from a height of 125 feet, estimate how long it will take to reach the ground. (Example 4)

_____

Order each set of numbers from least to greatest.

12. $\left\{7, 9, \sqrt{50}, \sqrt{85}\right\}$ _____

13. $\left\{\sqrt[3]{105}, 7, 5, \sqrt{38}\right\}$ _____

14. **MP Persevere with Problems** Amanda purchased a storage cube that has a volume of 4 cubic feet. She wants to put it on a bookshelf that is 12 inches tall. Will the cube fit? Explain. _____

15. Without a calculator, determine which is greater, $\sqrt{94}$ or 10. Explain your reasoning. _____

 H.O.T. Problems Higher Order Thinking

16. **MP Persevere with Problems** Find two numbers that have square roots between 7 and 8. One number should have a square root closer to 7 and the other number should have a square root closer to 8. Justify your answer.

_____

_____

_____

17. **MP Find the Error** Jasmine is estimating $\sqrt{200}$. Find her mistake and correct it. _____

_____

_____

_____

_____

$\sqrt{200} \approx 100$

18. **MP Construct an Argument** If $x^4 = y$, then x is the fourth root of y. Explain how to estimate the fourth root of 30. Find the fourth root of 30 to the nearest whole number.

_____

_____

19. **MP Reason Inductively** Suppose x is a number between 1 and 10 and y is a number between 10 and 20. Determine whether the statement below is always, sometimes or never true. Explain your reasoning.

$$\sqrt{x} > \sqrt[3]{y}$$

_____

_____

Extra Practice

Estimate to the nearest integer.

20. $\sqrt{44} \approx$ _7_____

$$36 < 44 \quad < 49$$
$$6^2 < 44 \quad < 7^2$$
$$\sqrt{6^2} < \sqrt{44} < \sqrt{7^2}$$
$\sqrt{44}$ is closer to $\sqrt{49}$ or 7.

21. $\sqrt[3]{199} \approx$ _6_____

$$125 < \quad 199 < 216$$
$$5^3 \quad < \sqrt[3]{199} < 6^3$$
$$\sqrt{5^3} < \sqrt[3]{199} < \sqrt{6^3}$$
$\sqrt[3]{199}$ is closer to $\sqrt{216}$ or 6.

22. $\sqrt{125} \approx$ _____

23. $\sqrt{23.5} \approx$ _____

24. $\sqrt[3]{59} \approx$ _____

25. $\sqrt[3]{430} \approx$ _____

Estimate the solution of each equation to the nearest integer.

26. $y^2 = 55$

27. $d^2 = 95$

28. $p^2 = 6.8$

The volume of each cube is given. Estimate the side length of the cube to the nearest integer. Use the formula $V = s^3$.

29. _____

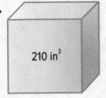

210 in³

30. _____

520 cm³

31. **MP** **Use Math Tools** Jacob is buying a bag of grass seed. The two-pound bag will cover 1,000 square feet of lawn. Estimate the side length of the largest square Jacob could seed if he purchases 5 bags.

_____

32. The radius of a circle with area A can be approximated using the formula

$r = \sqrt{\dfrac{A}{3}}$. Use a number line to estimate the radius of each circular swimming pool to the nearest integer.

Pool 1: $A = 240$ ft^2

$r \approx$ [   ]

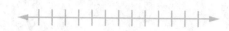

Pool 2: $A = 105$ ft^2

$r \approx$ [   ]

Pool 3: $A = 198$ ft^2

$r \approx$ [   ]

33. After an accident, officials use the formula $s = \sqrt{24m}$ to estimate the speed a car was traveling based on the length of the car's skid marks. In the formula, s represents the speed in miles per hour and m is the length of the skid marks in feet. If a car leaves a skid mark of 50 feet, what was its approximate speed?

[   ]

 Common Core Spiral Review

Write each of the following as a fraction in simplest form. 7.NS.2.d

34. $-36 =$ _____

35. $1.7 =$ _____

36. $-0.048 =$ _____

37. $98\% =$ _____

38. Of the 150 students in Mr. Bacon's classes, 16% play soccer, $\dfrac{9}{25}$ play basketball, 3^3 play football and 14 do not play a sport at all. Write the number of students in order from least to greatest. 6.NS.7

_____

Compare Real Numbers

 Real-World Link

Sports Major League baseball has rules for the dimensions of the baseball diamond. A model of the diamond is shown.

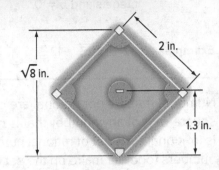

$\sqrt{8}$ in.
2 in.
1.3 in.

1. On the model, the distance from the pitching mound to home plate is 1.3 inches. Is 1.3 a rational number? Explain.

2. On the model, the distance from first base to second base is 2 inches. Is 2 a rational number? Explain.

3. The distance from home plate to second base is $\sqrt{8}$ inches. Using a calculator, find $\sqrt{8}$. Does it appear to terminate or repeat?

4. To determine if the number terminates, on your calculator, multiply your answer to $\sqrt{8}$ by itself. Do not use the x^2 button.

   Is the answer 8? _____

5. Based on your results, can you classify $\sqrt{8}$ as a rational number? Explain.

 Essential Question

WHY is it helpful to write numbers in different ways?

 Vocabulary

irrational number
real number

 Common Core State Standards

Content Standards
8.NS.1, 8.NS.2, 8.EE.2

MP **Mathematical Practices**
1, 3, 4, 6

Which MP Mathematical Practices did you use?
Shade the circle(s) that applies.

① Persevere with Problems
② Reason Abstractly
③ Construct an Argument
④ Model with Mathematics
⑤ Use Math Tools
⑥ Attend to Precision
⑦ Make Use of Structure
⑧ Use Repeated Reasoning

Real Numbers

Words

Rational Number

A rational number is a number that can be expressed as the ratio $\frac{a}{b}$, where a and b are integers and $b \neq 0$.

Irrational Number

An **irrational number** is a number that *cannot* be expressed as the ratio $\frac{a}{b}$, where a and b are integers and $b \neq 0$.

Examples $-2, 5, 3.\overline{76}, -12\frac{7}{8}$ $\sqrt{2} \approx 1.414213562...$

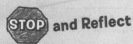
STOP and Reflect

Explain below how you know that $\sqrt{2}$ is an irrational number.

Numbers that are not rational are called irrational numbers. The square root of any number that is not a perfect square number is irrational. The set of rational numbers and the set of irrational numbers together make up the set of **real numbers**. Study the Venn diagram below.

Real Numbers

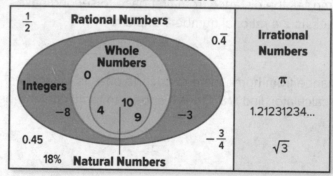

Examples

Tutor

Name all sets of numbers to which each real number belongs.

1. 0.2525...   The decimal ends in a repeating pattern. It is a rational number because it is equivalent to $\frac{25}{99}$.

2. $\sqrt{36}$ Since $\sqrt{36} = 6$, it is a natural number, a whole number, an integer, and a rational number.

Show your work.

a. _____

3. $-\sqrt{7}$ $-\sqrt{7} \approx -2.645751311...$ The decimal does not terminate nor repeat, so it is an irrational number.

b. _____

Got it? Do these problems to find out.

c. _____

a. $\sqrt{10}$ b. $-2\frac{2}{5}$ c. $\sqrt{100}$

Compare and Order Real Numbers

You can compare and order real numbers by writing them in the same notation. Write the numbers in decimal notation before comparing or ordering them.

Examples

Fill in each ◯ with <, >, or = to make a true statement.

4. $\sqrt{7}$ ◯ $2\frac{2}{3}$

$\sqrt{7} \approx 2.645751311...$

$2\frac{2}{3} = 2.666666666...$

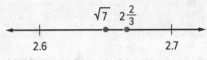

Since 2.645751311... is less than 2.66666666..., $\sqrt{7} < 2\frac{2}{3}$.

5. 15.7% ◯ $\sqrt{0.02}$

$15.7\% = 0.157$

$\sqrt{0.02} \approx 0.141$

Since 0.157 is greater than 0.141, $15.7\% > \sqrt{0.02}$.

6. Order the set $\left\{\sqrt{30}, 6, 5\frac{4}{5}, 5.3\overline{6}\right\}$ from least to greatest. Verify your answer by graphing on a number line.

Write each number as a decimal. Then order the decimals.

$\sqrt{30} \approx 5.48$

$6 = 6.00$

$5\frac{4}{5} = 5.80$

$5.3\overline{6} \approx 5.37$

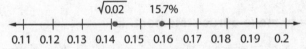

From least to greatest, the order is $5.3\overline{6}$, $\sqrt{30}$, $5\frac{4}{5}$, and 6.

Got it? Do these problems to find out.

d. $\sqrt{11}$ ◯ $3\frac{1}{3}$ **e.** $\sqrt{17}$ ◯ 4.03   **f.** $\sqrt{6.25}$ ◯ 250%

g. Order the set $\left\{-7, -\sqrt{60}, -7\frac{7}{10}, -\frac{66}{9}\right\}$ from least to greatest. Verify your answer by graphing on the number line below.

g. _____

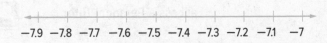

 Real World

Example

 Watch Tutor

7. On a clear day, the number of miles a person can see to the horizon is about 1.23 times the square root of his or her distance from the ground in feet. Suppose Frida is at the Empire State Building observation deck at 1,250 feet and Kia is at the Freedom Tower observation deck at 1,362 feet. How much farther can Kia see than Frida?

Use a calculator to approximate the distance each person can see.

Frida: $1.23 \cdot \sqrt{1,250} \approx 43.49$ Kia: $1.23 \cdot \sqrt{1,362} \approx 45.39$

Kia can see $45.39 - 43.49$ or 1.90 miles farther than Frida.

Guided Practice

Check ✓

Name all sets of numbers to which each real number belongs. (Examples 1–3)

1. 0.050505... _____

 Show your work.

2. $-\sqrt{64}$ _____

3. $\sqrt{17}$ _____

Fill in each ◯ with <, >, or = to make a true statement. (Examples 4 and 5)

4. $\sqrt{15}$ ◯ 3.5

5. $\sqrt{2.25}$ ◯ 150%

6. $\sqrt{6.2}$ ◯ $2.\overline{4}$

7. Order the set $\{\sqrt{5}, 220\%, 2.25, 2.\overline{2}\}$ from least to greatest. Verify your answer by graphing on a number line. (Example 6)

2.19 2.2 2.21 2.22 2.23 2.24 2.25 2.26 2.27 2.28 2.29 2.3

8. The formula $A = \sqrt{s(s-a)(s-b)(s-c)}$ can be used to find the area A of a triangle. The variables a, b, and c are the side measures and s is one half the perimeter. Use the formula to find the area of a triangle with side lengths of 7 centimeters, 9 centimeters, and

10 centimeters. (Example 7) _____

9. **Building on the Essential Question** How are real numbers different from irrational numbers?

_____

_____

_____

Rate Yourself!

How well do you understand real numbers? Circle the image that applies.

☀ Clear   ⛅ Somewhat Clear   ☁ Not So Clear

For more help, go online to access a Personal Tutor. Tutor

Independent Practice

Go online for Step-by-Step Solutions

Name all sets of numbers to which each real number belongs. (Examples 1–3)

1. $\frac{2}{3}$ _____

 Show your work.

2. $-\sqrt{20}$ _____

3. $7.\overline{2}$ _____

4. $\frac{12}{4}$ _____

Fill in each ◯ **with <, >, or = to make a true statement.** (Examples 4 and 5)

5. $\sqrt{10}$ ◯ 3.2

6. $5\frac{1}{6}$ ◯ $5.1\overline{6}$

7 $2.\overline{21}$ ◯ $\sqrt{5.2}$

Order each set of numbers from least to greatest. Verify your answer by graphing on a number line. (Example 6)

8. $\left\{-415\%, -\sqrt{17}, -4.\overline{1}, -4.08\right\}$

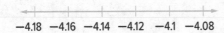

_____

9 $\left\{\sqrt{5}, \sqrt{6}, 2.5, 2.55, \frac{7}{3}\right\}$

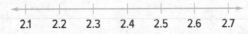

_____

10. The equation $s = \sqrt{30fd}$ can be used to find a car's speed s in miles per hour given the length d in feet of a skid mark and the friction factor f of the road. Police measured a skid mark of 90 feet on a dry concrete road. If the speed limit is 35 mph, was the car speeding? Explain. (Example 7)

Friction Factor		
Road	Concrete	Tar
Wet	0.4	0.5
Dry	0.8	1.0

_____

11. The surface area in square meters of the human body can be found using the expression $\sqrt{\frac{hm}{3,600}}$ where h is the height in centimeters and m is the mass in kilograms. Find the surface area of a 15-year-old boy with a height of 183 centimeters and a mass of 74 kilograms. (Example 7)

_____

12. **MP Be Precise** Write a brief description and give an example of each type of number in the graphic organizer shown.

natural	whole	integer	rational	irrational

Use estimation to fill in each ◯ with <, >, or = to make a true statement.

13. 3π ◯ $\sqrt{78}$

14. π^2 ◯ $3 \cdot \sqrt{15}$

15. $\sqrt{980}$ ◯ $4\pi^2$

H.O.T. Problems Higher Order Thinking

16. **MP Use a Counterexample** Give a counterexample for the statement *All square roots are irrational numbers.* Explain your reasoning.

_____

MP Persevere with Problems Tell whether the following statements are *always*, *sometimes*, or *never* true. If a statement is not always true, explain.

17. Integers are rational numbers. _____

_____

18. Rational numbers are integers. _____

_____

19. The product of a non-zero rational number and an irrational number is

irrational. _____

_____

20. **MP Model with Mathematics** Identify two numbers, one rational number and one irrational number, that are between 1.4 and 1.6. Include the decimal approximation of the irrational number to the nearest hundredth.

_____

Extra Practice

21. Name all sets of real numbers to which
$\sqrt{10}$ belongs. _irrational_

Homework
Help

$\sqrt{10} \approx 3.16227766...$ Since the decimal does
not terminate nor repeat, it is an irrational
number.

22. Fill in ⭕ with <, >, or = to make
$5.1\overline{5}$ ⟩ $\sqrt{26}$ a true statement.

Write each number as a decimal.
$5.1\overline{5} = 5.155555...$
$\sqrt{26} \approx 5.099019...$
Since 5.155555... is greater than 5.099019...,
$5.1\overline{5} > \sqrt{26}$.

Name all sets of numbers to which each real number belongs.

23. 14

24. $-\sqrt{16}$

25. $-\sqrt[3]{90}$

Fill in each ⭕ **with <, >, or = to make a true statement.**

26. $\sqrt{12}$ ⭕ 3.5

27. $6\frac{1}{3}$ ⭕ $\sqrt[3]{240}$

28. 240% ⭕ $\sqrt{5.76}$

29. About how much greater is the perimeter of a square with area 250 square
meters than a square with an area of 125 square meters?

_____

30. **MP** **Persevere with Problems** In the sequence 4, 12, ■, 108, 324, the
missing number can be found by simplifying $\sqrt{ab}$ where a and b are the
numbers on either side of the missing number. Find the missing number.

_____

Fill in each ⭕ **with <, >, or = to make a true statement.**

31. $3 + \sqrt{7}$ ⭕ 6

32. $4 - \sqrt{10}$ ⭕ $\sqrt{2}$

33. 13 ⭕ $8 + \sqrt{20}$

34. Mr. Rodgers gave his students a test that was worth 100 points. Erika earned an 84%, Joe earned $\frac{5}{6}$ of the total points, Malcolm earned $\sqrt{7225}$ points, and Stephanie earned $\frac{83}{100}$ points. Plot points on the number line to represent each students' score.

Number of Points

Which student earned the most points?

35. The diagonal of a rectangular room is $\sqrt{289}$ feet long. To which sets of numbers does $\sqrt{289}$ belong? Select all that apply.

- ☐ real
- ☐ rational
- ☐ whole
- ☐ integer
- ☐ irrational
- ☐ natural

Common Core Spiral Review

36. Order the set $\{7, \sqrt{53}, \sqrt{32}, 6\}$ from least to greatest. **8.EE.2**

_____

Solve each equation. 8.EE.2

37. $t^2 = 25$ _____

38. $y^2 = \frac{1}{49}$ _____

39. $0.64 = a^2$ _____

Evaluate each expression. Express the result in scientific notation. 8.EE.4

40. $(7.2 \times 10^4)(1.1 \times 10^{-6}) =$ _____

41. $(3.6 \times 10^3) + (5.7 \times 10^5) =$ _____

42. The table shows the approximate population of several countries. Order the countries from the greatest population to the least population. **8.EE.4**

_____

Country	Population
China	1.3×10^9
India	1.2×10^9
Indonesia	2.3×10^8
United States	3.1×10^8

21ST CENTURY CAREER
in Engineering

Robotics Engineer

Are you mechanically inclined? Do you like to find new ways to solve problems? If so, a career as a robotics engineer is something you should consider. Robotics engineers design and build robots to perform tasks that are difficult, dangerous, or tedious for humans. For example, a robotic insect was developed based on a real insect. Its purpose was to travel over water surfaces, take measurements, and monitor water quality.

Is This the Career for You?

Are you interested in a career as a robotics engineer? Take some of the following courses in high school.

- ◆ Calculus
- ◆ Electro-Mechanical Systems
- ◆ Fundamentals of Robotics
- ◆ Physics

Turn the page to find out how math relates to a career in Engineering.

Hill Education   (t)Courtesy of Georgia Institute of Technology/Rob Felt; (b)PhotoLink/Getty Images

MP Relying on Robots

Use the information in the table to solve each problem.

1. Write the mass of the robot in standard form. _____

2. Write the length of the robot in scientific notation. _____

3. Write the leg diameter of the robot in scientific notation. _____

4. What is the mass in milligrams? Write in standard form. _____

5. Real insects called water striders can travel 8.3 times faster than the robot. Write the speed of water striders in scientific notation.

_____

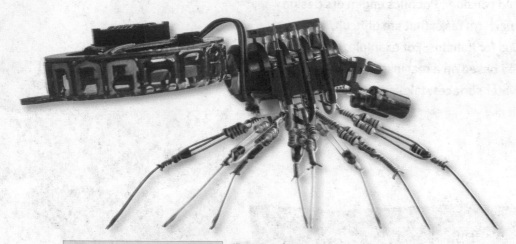

Robotic Insect Characteristics	
Mass	3.5×10^{-4} kg
Length	0.09 m
Leg Diameter	0.2 mm
Speed	180 mm/s

MP Career Project

It's time to update your career portfolio! Investigate the education and training requirements for a career in robotics engineering.

_____

_____

_____

_____

_____

_____

What skills would you need to improve to succeed in this career?

- _____

- _____

- _____

- _____

- _____

Chapter Review ✓

Vocabulary Check

Complete the crossword puzzle using the vocabulary list at the beginning of the chapter.

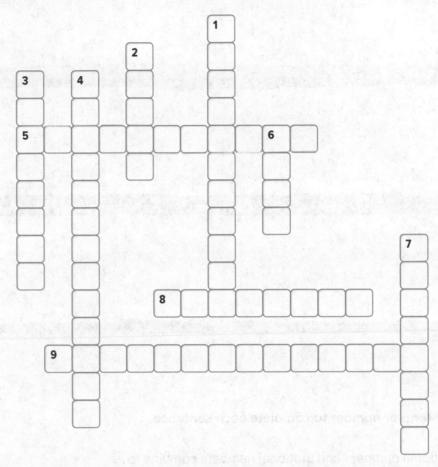

Across

5. a rational number whose cube root is a whole number
8. a number, a variable, or a product of a number and one or more variables
9. numbers that can be written as a comparison of two integers, expressed as a fraction

Down

1. the symbol used to indicate a positive square root
2. a product of repeated factors using a base and exponent
3. this tells how many times a number is used as a factor
4. a rational number whose square root is a whole number
6. in a power, the number that is the common factor
7. one of a number's three equal factors

Use Your FOLDABLES

Use your Foldable to help review the chapter.

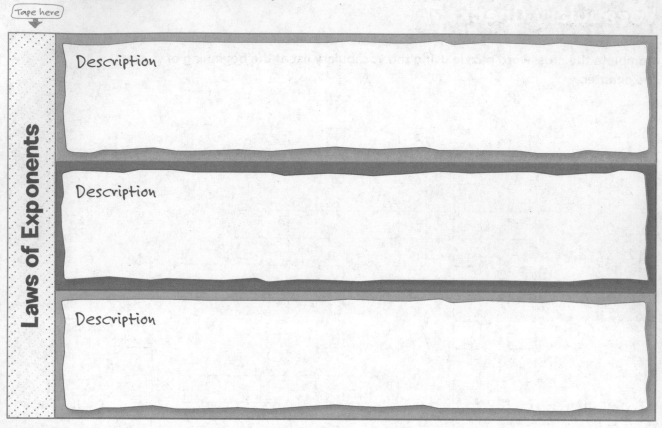

Tape here

Laws of Exponents

Description

Description

Description

Got it?

(Circle) the correct term or number to complete each sentence.

1. The sets of rational numbers and irrational numbers combine to make the (whole, real) numbers.

2. The product of $3a^2b$ and $-3a^2b$ is $(-9a^4b^2, a^4b)$.

3. You would use the (Power of a Product rule, Product of Powers rule) to simplify the expression $(p^2r)^4$.

4. The expression $\dfrac{6^2 \cdot 2^6 \cdot 8^4}{6 \cdot 2^3 \cdot 8^3}$ is equal to (384, 288).

5. Another way to write $(9^4)^7$ is $(9^{11}, 9^{28})$.

6. 3^{-4} is equal to $\left(-81, \dfrac{1}{81}\right)$.

7. Scientific notation is when a number is written as a product of a power of 10 and a factor greater than or equal to 1 and (less than, less than or equal to) 10.

Power Up! Performance Task

Planetary Play

The table shows some of the planets' approximate distances from the sun.

Planet	Approximate Distance from the Sun (km)
Mercury	5.8×10^7
Venus	1.1×10^8
Earth	1.5×10^8
Mars	2.3×10^8
Jupiter	7.8×10^8
Saturn	1.4×10^9
Neptune	4.5×10^9

Write your answers on another piece of paper. Show all of your work to receive full credit.

Part A
How much farther is Earth from the Sun than it is from Venus? How many times further is Neptune from the Sun than Mercury is from the Sun? Explain.

Part B
It takes Mercury approximately 2.4×10^{-1} Earth years to orbit the Sun. Write this in standard form and in fraction form.

Part C
The volume of Saturn is 8.27×10^{14}. This is about 766 times the volume of Earth. What is the approximate volume of Earth? Show your work.

Part D
In her science class, Nadia needs to make a model showing the planets' relative distance from the Sun. She wants to use centimeter grid paper to create her model. If 1 centimeter on the model represents 100,000,000 kilometers, write a proportion Nadia could use to create her model on the grid paper. Use the proportion to find the distance for each planet on the model. Then use a ruler and centimeter grid paper to draw a model similar to Nadia's.

Reflect

 Answering the Essential Question

Use what you learned about numbers to complete the graphic organizer.
For each category, describe why you would use that form for the number
$35,036\frac{1}{3}$. Then write the number in that form. If you would not use the
number in that form, explain why.

Decimal	Power

WHY is it helpful to write numbers in different ways?

Fraction	Scientific Notation

 Answer the Essential Question. WHY is it helpful to write numbers in
different ways?

UNIT PROJECT

Watch ▶ **Music to My Ears** When you listen to music, you may not be aware of the math used to create it. In this project you will:

- **Collaborate** with your classmates as you research the connections between math and music.
- **Share** the results of your research in a creative way.
- **ℯ Reflect** on how mathematical ideas can be represented.

By the end of this project, you just might be ready to write a hit song!

Collaborate

ᕮ Go Online Work with your group to research and complete each activity. You will use your results in the Share section on the following page.

1. Choose a song on a CD or on your MP3 player. Listen to the song and describe the beats or rhythm using repeating numbers. For example, a song may have a rhythm that can be described by 1-2-3-1-2-3-... .

2. Research and describe the different types of musical notes. Make sure to use rational numbers. Include a drawing of each note on sheet music.

3. Research Pythagoras' findings about music, notes and frequency, and harmony. Write a few paragraphs about what you found and create a list of the types of numbers you find in your research.

4. Describe the Fibonacci Sequence. Then give some examples of how Fibonacci numbers are found in music.

5. Find the digital music sales in a recent year. Write this number in both standard form and scientific notation. Then compare the digital music sales to CD music sales for the same year. Create a display to show what you find.

With your group, decide on a way to share what you have learned about math and music. Some suggestions are listed below, but you can also think of other creative ways to present your information. Remember to show how you used mathematics to complete each of the activities in this project!

connect with Health

Health Literacy Many studies have been done that show a positive connection between music and good health. Research the Internet to find information about one such study.

- Create your own short piece of music based on your knowledge of notes and frequency. Make a recording of the music and explain how it is harmonious.
- Use presentation software to demonstrate some ways math and music are connected.

Check out the note on the right to connect this project with other subjects.

Reflect

On Your Own

6. **Answer the Essential Question** HOW can mathematical ideas be represented?

    a. How did you use what you learned about real numbers in this chapter to represent mathematical ideas in this project?

    _____

    _____

    _____

    b. In this project, you discovered how mathematical ideas are represented in music. Explain how mathematical ideas are represented in other parts of culture.

    _____

    _____

    _____

UNIT 2

Expressions and Equations

CCSS

Essential Question

HOW can you communicate mathematical ideas effectively?

Chapter 2
Equations in One Variable

Linear equations in one variable can have one solution, infinitely many solutions, or no solutions. In this chapter, you will write and solve two-step equations and solve equations with variables on both sides.

Chapter 3
Equations in Two Variables

In a proportional relationship, the unit rate is the slope of the graph. In this chapter, you will graph equations of the form $y = mx$ and $y = mx + b$. You will then solve systems of equations algebraically and by graphing.

Copyright © McGraw-Hill Education   (l)George Doyle/Stockbyte/Getty Images, Steven P. Lynch; (t)DreamPictures/Blend Images LLC; (c)Stocktrek Images/Getty Images

Collaborate

Unit Project Preview

Watch

Web Design 101 A Web page is a useful way of presenting a summary of facts and statistics about a subject.

To design a Web page, you must first collect all the information that you want to include on the page. You will also need to decide how to balance the text and graphics. This will ensure that your Web page not only looks good but is functional too.

At the end of Chapter 3, you'll complete a project in which you will plan the design of a Web page about your favorite insect or other animal. But for now, it's time to do an activity in your book. Choose a subject that interests you. In the space provided, make an outline of all the items you would include if you were to design a Web page about that subject.

My Web page about

_____

would include:

Chapter 2
Equations in One Variable

Essential Question

WHAT is equivalence?

Common Core State Standards

Content Standards
8.EE.7, 8.EE.7a, 8.EE.7b

MP Mathematical Practices
1, 2, 3, 4, 5, 7

Math in the Real World

Tips Maria and her family had salads and dinner at a local restaurant. Her mother wants to leave an 18% tip. The proportion $\dfrac{p}{\$35.60} = \dfrac{18}{100}$ can be used to find the amount of tip she should leave.

Use the proportion to find the amount of tip Maria's mother should leave. Then find the total amount.

Pizzeria

Date:
Card Type:
Acct. No.:
Exp. Date:
Check:
Table:

Subtotal: **$35.60**

Tip: _ _ _ _ _ _ _

Total: _ _ _ _ _ _ _

FOLDABLES®
Study Organizer

1 Cut out the Foldable on page FL5 of this book.

2 Place your Foldable on page 164.

3 Use the Foldable throughout this chapter to help you learn about solving equations.

Vocabulary

coefficient	multiplicative inverse	properties
identity	null set	two-step equation

Study Skill: Writing Math

Justify Your Answer When you justify your answer, you give *reasons* why your answer is correct.

The different plans an online movie rental company offers are shown. Mariah wants to purchase the 2 DVDs at-a-time plan. This month, the plans are advertised at $\frac{1}{4}$ off. If she has $10.00 to spend, does she have enough? Justify your answer.

Online DVD Rental	
Plan	Monthly Price ($)
3 DVDs at-a-time	14.50
2 DVDs at-a-time	12.00
1 DVD at-a-time	8.00

Step 1 **Solve the problem.**	Find the discount. $\frac{1}{4}$ of $12 = 3$ The discount is $3.00. Find the discounted price. $12 − $3 = $9
Step 2 **Answer the question.**	Mariah does have enough money.
Step 3 **Justify your answer.** Always write complete sentences.	Mariah has enough money because $9.00 is less than $10.00.

You can buy 3 used CDs at The Music Shoppe for $12.99, or you can buy 5 for $19.99 at Quality Sounds. Which is the better buy? Justify your answer.

Step 1 **Solve the problem.**	
Step 2 **Answer the question.**	
Step 3 **Justify your answer.** Always write complete sentences.	

What Do You Already Know?

Read each statement. Decide whether you agree (A) or disagree (D). Place a checkmark in the appropriate column and then justify your reasoning.

Equations in One Variable			
Statement	A	D	Why?
To solve an equation with a fractional coefficient, multiply each side of the equation by the multiplicative inverse of the fraction.			
The first step to solve $-2x + 9 = 6$ is to divide both sides by -2.			
If an equation has variables on both sides of the equals sign, the first thing you should do is get the variables on the same side of the equation.			
Equations can have no solution, one solution, or infinitely many solutions.			
You would use the Associative Property first to solve the equation $6(x - 3) + 10 = 2(3x + 4)$.			
An equation that is true for every value of the variable is called an identity.			

When Will You Use This?

Here are a few examples of how equations are used in the real world.

Activity 1 Do you or your parents have a texting plan? If so, how much does it cost per text or per month? Ask your parents to help you research different texting plans. Then compare and contrast each plan.

Activity 2 Go online at **connectED.mcgraw-hill.com** to read the graphic novel *Texting, Texting, 1, 2, 3*. What is each person's cell phone plan?

Roberto and Jacob in

Texting, Texting, 1, 2, 3

Hm. I've sent 54 so far. Wonder how many more we've each got...

Try the Quick Check below.
Or, take the Online Readiness Quiz.

 Check ✓

Example 1

Solve $44 = k - 7$.

$44 = k - 7$	Write the equation.
$+7 = +7$	Addition Property of Equality
$51 = k$	

Example 2

Solve $18m = -360$.

$18m = -360$	Write the equation.
$\dfrac{18m}{18} = \dfrac{-360}{18}$	Division Property of Equality
$m = -20$	Simplify.

Quick Check

One-Step Equations **Solve each equation. Check your solution.**

1. $n + 8 = -9$

2. $4 = p + 19$

3. $-4 + a = 15$

Show your work.

4. $3c = -18$

5. $-42 = -6b$

6. $\dfrac{w}{4} = -8$

7. Barry has 18 more marbles than Heidi. If Barry has 92 marbles, write and solve an equation to determine the number of marbles Heidi has.

_____

How Did You Do?

Which problems did you answer correctly in the Quick Check?
Shade those exercise numbers below.

 1 2 3 4 5 6 7

Solve Equations with Rational Coefficients

Essential Question

WHAT is equivalence?

Vocab
$\frac{a}{b_c}$ **Vocabulary**

multiplicative inverse
coefficient

CCSS **Common Core State Standards**

Content Standards
8.EE.7, 8.EE.7a, 8.EE.7b
MP **Mathematical Practices**
1, 3, 4, 7

Vocabulary Start-Up

Two numbers with a product of 1, such as $\frac{3}{4}$ and $\frac{4}{3}$, are called reciprocals or **multiplicative inverses**.

Complete the graphic organizer.

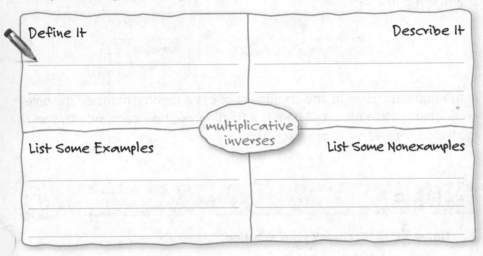

Define It

Describe It

multiplicative inverses

List Some Examples

List Some Nonexamples

Describe how a multiplicative inverse is used in division of fractions.

Real-World Link

How can the action of the motorcyclist in the photo help you remember what the multiplicative inverse is? _____

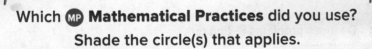

Yikes!

Which MP **Mathematical Practices did you use?**
Shade the circle(s) that applies.

① Persevere with Problems

⑤ Use Math Tools

② Reason Abstractly

⑥ Attend to Precision

③ Construct an Argument

⑦ Make Use of Structure

④ Model with Mathematics

⑧ Use Repeated Reasoning

Inverse Property of Multiplication

Words   The product of a number and its multiplicative inverse is 1.

Numbers $\frac{7}{8} \times \frac{8}{7} = 1$ $-\frac{3}{2} \times -\frac{2}{3} = 1$

Symbols $\frac{a}{b} \cdot \frac{b}{a} = 1$, where a and $b \neq 0$

STOP and Reflect

What is the multiplicative inverse of $-\frac{3}{2}$?

The numerical factor of a term that contains a variable is called the **coefficient** of the variable.

coefficient ┈┈▶ **3x** ◀┈┈ variable

In the equation $\frac{3}{4}c = 18$, the coefficient of c is a rational number. To solve an equation when the coefficient is a fraction, multiply each side by the multiplicative inverse of the fraction.

Example

1. Solve $\frac{3}{4}c = 18$. **Check your solution.**

$\frac{3}{4}c = 18$ Write the equation.

$\left(\frac{4}{3}\right) \cdot \frac{3}{4}c = \left(\frac{4}{3}\right) \cdot 18$ Multiply each side by the multiplicative inverse of $\frac{3}{4}$, $\frac{4}{3}$.

$\frac{\cancel{4}^{1}}{3} \cdot \frac{\cancel{3}^{1}}{\cancel{4}_{1}}c = \frac{4}{3} \cdot \frac{\cancel{18}^{6}}{1}$ Write 18 as $\frac{18}{1}$. Divide by common factors.

$c = 24$ Simplify.

Check $\frac{3}{4}c = 18$ Write the original equation.

$\frac{3}{4}(24) \stackrel{?}{=} 18$ Replace c with 24.

$\frac{3}{\cancel{4}_{1}}\left(\frac{\cancel{24}^{6}}{1}\right) \stackrel{?}{=} 18$ Write 24 as $\frac{24}{1}$. Divide by common factors.

$18 = 18 \ \checkmark$ This sentence is true.

Got it? Do these problems to find out.

a. _____

b. _____

c. _____

d. _____

Show your work.

   a. $\frac{1}{5}x = 12$ **b.** $-\frac{2}{9}d = 4$

   c. $15 = \frac{5}{3}n$ **d.** $-24 = -\frac{6}{7}p$

Example

2. Solve $1\frac{1}{2}s = 16\frac{1}{2}$. Check your solution.

$$1\frac{1}{2}s = 16\frac{1}{2}$$ Write the equation.

$$\frac{3}{2}s = \frac{33}{2}$$ Rename $1\frac{1}{2}$ as $\frac{3}{2}$ and $16\frac{1}{2}$ as $\frac{33}{2}$.

$$\left(\frac{2}{3}\right) \cdot \frac{3}{2}s = \left(\frac{2}{3}\right) \cdot \frac{33}{2}$$ Multiply each side by the multiplicative inverse of $\frac{3}{2}$, $\frac{2}{3}$.

$$\frac{\overset{1}{\cancel{2}}}{\underset{1}{\cancel{3}}} \cdot \frac{\overset{1}{\cancel{3}}}{\underset{1}{\cancel{2}}}s = \frac{\overset{1}{\cancel{2}}}{\underset{1}{\cancel{3}}} \cdot \frac{\overset{11}{\cancel{33}}}{\underset{1}{\cancel{2}}}$$ Divide by common factors.

$$s = 11$$ Simplify.

Got it? Do these problems to find out.

d. $4\frac{1}{6} = 3\frac{1}{3}c$ **e.** $-9\frac{5}{8}w = 108$ **f.** $1\frac{7}{8}y = 4\frac{1}{2}$

Show your work.

d. _____

e. _____

f. _____

Solve Equations with Decimal Coefficients

In the equation $3.15 = 0.45n$ the coefficient of n is a decimal. To solve an equation with a decimal coefficient, divide each side of the equation by the coefficient.

Quick Review
Division

$$0.45\overline{)3.15}$$ with quotient 7, $-3\,15$, remainder 0

Example

3. Solve $3.15 = 0.45n$. Check your solution.

$$3.15 = 0.45n$$ Write the equation.

$$\frac{3.15}{0.45} = \frac{0.45n}{0.45}$$ Division Property of Equality

$$7 = n$$ Simplify.

Check $3.15 = 0.45n$ Write the original equation.

$$3.15 = 0.45(7)$$ Replace n with 7.

$$3.15 = 3.15 \checkmark$$ The sentence is true.

g. _____

Got it? Do these problems to find out.

h. _____

g. $4.9 = 0.7t$ **h.** $-1.4m = 2.1$ **i.** $-5.6k = -12.88$

i. _____

Example

 Tutor

4. Latoya's softball team won 75%, or 18, of its games. Define a variable. Then write and solve an equation to determine the number of games the team played.

Latoya's softball team won 18 games, which was 75% of the games played. Let n represent the number of games played. Write and solve an equation.

$$0.75n = 18 \qquad \text{Write the equation. Write 75\% as 0.75.}$$

$$\frac{0.75n}{0.75} = \frac{18}{0.75} \qquad \text{Division Property of Equality}$$

$$n = 24 \qquad \text{Simplify.}$$

Latoya's softball team played 24 games.

Guided Practice

 Check ✓

Solve each equation. Check your solution. (Examples 1–3)

1. $60 = \frac{3}{4}p$

2. $-\frac{27}{25}x = -\frac{9}{5}$

3. $-2.7t = 810$

 Show your work.

4. Paula has read 70% of the total pages in a book she is reading for English class. Paula has read 84 pages. Define a variable. Then write and solve an equation to determine how many pages are in the book. (Example 4)

_____

5. ⓔ **Building on the Essential Question** How is the multiplicative inverse used to solve an equation that has a rational coefficient?

_____

_____

_____

_____

Rate Yourself!

Are you ready to move on? Shade the section that applies.

I have a few questions. | I'm ready to move on.

I have a lot of questions.

For more help, go online to access a Personal Tutor. Tutor

Independent Practice

Go online for Step-by-Step Solutions

Solve each equation. Check your solution. (Examples 1–3)

1. $6 = \frac{1}{12}v$

Show your work.

2. $-\frac{2}{3}w = 60$

3 $-\frac{7}{8}k = -21$

4. $9.6 = 1.2b$

5. $0.75a = -9$

6. $-413.4 = -15.9n$

7. $3\frac{1}{10}s = 6\frac{1}{5}$

8. $2\frac{2}{9} = -\frac{4}{5}m$

9. $-2\frac{4}{5} = -3\frac{1}{2}n$

Define a variable. Then write and solve an equation for each situation. (Example 4)

10. The Parker family drove a total of 180 miles on their road trip. This distance is 1.5 times the distance they drove on the first day. How many miles did the Parker family drive on the first day?

11 José correctly answered 80% of the questions on a language arts quiz. If he answered 16 questions correctly, how many questions were on the language arts quiz?

12. **Financial Literacy** Demitrius deposited 60% of his paycheck into his savings account. What was the amount of his paycheck?

Savings Deposit Slip	
Demetrius Matthews	
Name	
Amount Deposited	$41.67

13.  **Identify Structure** Suppose the numbers, $1\frac{1}{3}$, 0.2, -5, $-\frac{1}{2}$, are each coefficients in separate equations. Choose whether you would solve the equation by multiplying each side by the multiplicative inverse of the coefficient or by dividing each side by the coefficient. Write the number in the appropriate space.

Multiplicative Inverse

Division

 H.O.T. Problems Higher Order Thinking

14.  **Model with Mathematics** Write a real-world problem that can be represented by the equation $\frac{3}{4}c = 21$. _____

_____

_____

 **Persevere with Problems** Determine whether each statement is *true* or *false*. **Explain your reasoning.**

15. The product of a fraction and its multiplicative inverse is 1.

_____

_____

16. To solve an equation with a coefficient that is a fraction, divide each side of the equation by the reciprocal of the fraction. _____

_____

_____

17.  **Reason Inductively** Complete the statement: If $10 = \frac{1}{5}x$, then $x + 3 = \blacksquare$. Explain your reasoning. _____

_____

18.  **Justify Conclusions** Suppose your friend says he can solve $3x = 15$ by using the Multiplication Property of Equality. Is he correct? Justify your response. _____

_____

Name _____ My Homework _____

Extra Practice

Solve each equation. Check your solution.

19. $\frac{1}{2} = \frac{2}{5}z$

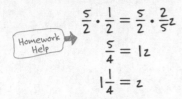

$$\frac{5}{2} \cdot \frac{1}{2} = \frac{5}{2} \cdot \frac{2}{5}z$$
$$\frac{5}{4} = 1z$$
$$1\frac{1}{4} = z$$

20. $-\frac{3}{4}t = 5$

21. $-\frac{2}{9}g = -\frac{7}{9}$

22. $0.6w = 0.48$

23. $-226.8 = 21.6y$

24. $-30 = 1.25c$

25. $1\frac{1}{2}x = 9\frac{9}{20}$

26. $-12\frac{2}{3} = -1\frac{1}{9}y$

27. $1\frac{5}{7} = 1\frac{13}{14}a$

28. One third of the bagels in a bakery are sesame bagels. There are 72 sesame bagels. Define a variable. Then write and solve an equation to find how many bagels there were in the bakery.

_____

29. **MP Find the Error** Sarah is solving the equation $-\frac{7}{8}x = 24$. Circle her mistake and correct it.

_____

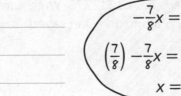

$$-\frac{7}{8}x = 24$$
$$\left(\frac{7}{8}\right) -\frac{7}{8}x = 24\left(\frac{7}{8}\right)$$
$$x = 21$$

Lesson 1 Solve Equations with Rational Coefficients **117**

30. Select the correct equation for each situation. Then solve each problem.

$\frac{x}{5} = 240$	$5x = 240$
$\frac{x}{240} = 0.5$	$0.05x = 240$
$0.5x = 240$	$\frac{5}{x} = 240$

  a. The Jansen family drove a total of 240 miles on their road trip. This distance is 5 times the distance they drove on the first day. How many miles did the family drive on the first day?

  Equation: [ ]   Solution: [ ]

  b. There are 240 students in Misty's school. This is 5% of the total students in the school district. How many students are there in the school district?

  Equation: [ ]   Solution: [ ]

31. The table shows how many miles Uyen has run this week.

Day	Monday	Tuesday	Wednesday	Thursday	Friday
Miles	6.5	2.9	4.2	5.5	3.1

The total distance this week is 1.5 times the distance that she ran last week.

How many miles did Uyen run last week? [ ]

CCSS Common Core Spiral Review

Solve each equation. Check your solution. **7.EE.4**

32. $w + 5 = -20$

33. $x - 17 = -32$

34. $t + 7.2 = 1.65$

35. $-0.4 = g - 4.9$

36. $y - \frac{2}{5} = 1\frac{3}{5}$

37. $-5\frac{1}{6} = 2\frac{1}{3} + p$

38. Financial Literacy Simone saved $65.35 more than her brother Dan and $37.50 less than her sister Carly. Carly saved $127.75. Write and solve equations to find how much money Simone and Dan saved. **6.EE.7**

Inquiry Lab

Solve Two-Step Equations

 Inquiry HOW does a bar diagram help you solve a real-world problem involving a two-step equation?

  Content Standards 8.EE.7, 8.EE.7a

MP Mathematical Practices 1, 2, 3, 4

Miranda bought two large postcards and four small postcards at a souvenir shop. Each small postcard costs $0.50. If Miranda spent $5.00 on postcards, what is the cost of one large postcard?

What do you know? _____

_____

What do you need to find? _____

Hands-On Activity

Step 1 The bar diagram represents the total number of postcards and the total cost. Label the missing parts.

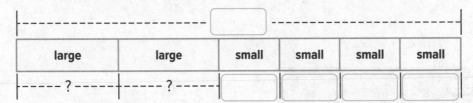

Step 2 Fill in the boxes to write an equation that represents the bar diagram. The cost of a large postcard is the unknown, so it is represented by the variable p.

$2p + \boxed{} = \boxed{}$

Step 3 Find the cost of the large postcards by working backward.

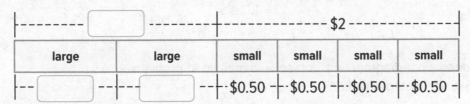

The cost of one large postcard is _____.

MP Reason Abstractly Work with a partner. Use a bar diagram to write and solve an equation for each exercise.

1. Brooke and two friends went to the movies and spent a total of $42. The movie tickets were $5 each and they each bought a popcorn combo. What is the cost of one popcorn combo?

Show your work.

2. Four medium postcards and 4 small postcards cost $5. What is the cost of one medium postcard?

$5							
medium	medium	medium	medium	small	small	small	small
				0.50	0.50	0.50	0.50

Create

On Your Own

3. **MP Model with Mathematics** Write and solve a word problem that could be represented by the bar diagram shown.

4. **Inquiry** HOW does a bar diagram help you solve a real-world problem involving a two-step equation?

Solve Two-Step Equations

Vocabulary Start-Up

Recall that in mathematics, **properties** are statements that are true for any number.

Complete the graphic organizer by matching the Property of Equality with the correct example.

Addition Property of Equality	$\frac{1}{2}x = 10$ $2 \cdot \frac{1}{2}x = 10 \cdot 2$
Division Property of Equality	$3x = 9$ $\frac{3x}{3} = \frac{9}{3}$
Multiplication Property of Equality	$x + 3 = 1$ $x + 3 - 3 = 1 - 3$
Subtraction Property of Equality	$x - 5 = 6$ $x - 5 + 5 = 6 + 5$

Essential Question

WHAT is equivalence?

Vocabulary

properties
two-step equation

Common Core State Standards

Content Standards
8.EE.7, 8.EE.7a, 8.EE.7b

MP Mathematical Practices
1, 2, 3, 4

Real-World Link

A property in science is a trait of matter that is always true under a given set of conditions. For example, pure water freezes at 0°C. How is the definition of *property* similar in science and math? _____

_____

_____

Which MP Mathematical Practices did you use?
Shade the circle(s) that applies.

① Persevere with Problems   ⑤ Use Math Tools

② Reason Abstractly   ⑥ Attend to Precision

③ Construct an Argument   ⑦ Make Use of Structure

④ Model with Mathematics   ⑧ Use Repeated Reasoning

Solve Two-Step Equations

A **two-step equation** contains two operations. In the equation $2x + 3 = 7$, x is multiplied by 2 and then 3 is added. To solve two-step equations, undo each operation in reverse order.

Example

1. Solve $2x + 3 = 7$.

> **Method 1** Use a model.

Remove three 1-tiles from each mat.

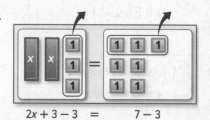

$$2x + 3 - 3 = 7 - 3$$

Separate the remaining tiles into 2 equal groups.

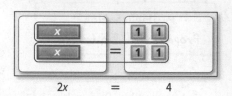

$$2x = 4$$

There are two 1-tiles in each group, so $x = 2$.

> **Method 2** Use symbols.

$$2x + 3 = 7 \qquad \text{Write the equation.}$$

$$\frac{-3 = -3}{2x \quad = 4} \qquad \text{Subtraction Property of Equality}$$

$$\frac{2x}{2} = \frac{4}{2} \qquad \text{Division Property of Equality}$$

$$x = 2 \qquad \text{Simplify.}$$

Using either method, the solution is 2.

Show your work.

a. _____

b. _____

Got it? Do these problems to find out.

a. $3x + 2 = 20$

b. $5 + 2n = -1$

Example

2. Solve $25 = \frac{1}{4}n - 3$.

$25 = \frac{1}{4}n - 3$	Write the equation.
$\underline{+3 = \quad +3}$	Addition Property of Equality
$28 = \frac{1}{4}n$	Simplify.
$4 \cdot 28 = 4 \cdot \frac{1}{4}n$	Multiplication Property of Equality
$112 = n$	

The solution is 112.

Got it? Do these problems to find out.

c. $-1 = \frac{1}{2}a + 9$ **d.** $\frac{2}{5}r - 5 = 7$

c. _____

d. _____

Example

3. Solve $6 - 3x = 21$.

$6 - 3x = 21$	Write the equation.
$6 + (-3x) = 21$	Rewrite the left side as addition.
$\underline{-6 \qquad = -6}$	Subtraction Property of Equality
$-3x = 15$	Simplify.
$\frac{-3x}{-3} = \frac{15}{-3}$	Division Property of Equality
$x = -5$	Simplify.

The solution is -5.

Check $6 - 3x = 21$	Write the equation.
$6 - 3(-5) \overset{?}{=} 21$	Replace x with -5.
$6 - (-15) \overset{?}{=} 21$	Multiply.
$6 + 15 \overset{?}{=} 21$	To subtract a negative number, add its opposite.
$21 = 21$ ✓	The sentence is true.

Common Error

A common mistake when solving the equation in Example 3 is to divide each side by 3 instead of -3. Since $6 - 3x = 6 + (-3x)$, the coefficient is -3.

Got it? Do these problems to find out.

e. $10 - \frac{2}{3}p = 52$ **f.** $-19 = -3x + 2$ **g.** $\frac{n}{-3} - 2 = -18$

e. _____

f. _____

g. _____

Example

4. **STEM**  Chicago's lowest recorded temperature in degrees Fahrenheit is $-27°$. Solve the equation $-27 = 1.8C + 32$ to convert to degrees Celsius.

$-27 = 1.8C + 32$	Write the equation.
$\underline{-32 = \qquad -32}$	Subtraction Property of Equality
$-59 = 1.8C$	Simplify.
$\dfrac{-59}{1.8} = \dfrac{1.8C}{1.8}$	Division Property of Equality
$-32.8 \approx C$	Simplify. Check the solution.

So, Chicago's lowest recorded temperature is about -32.8 degrees Celsius.

Guided Practice

Check ✓

Solve each equation. Check your solution. (Examples 1–3)

1. $6x + 5 = 29$

2. $3 - 5y = -37$

3. $\dfrac{2}{3}x - 5 = 7$

4. Cassidy went to the movies with some of her friends. The tickets cost $6.50 each, and they spent $17.50 on snacks. The total amount paid was $63.00. Solve the equation $63 = 6.50p + 17.50$ to determine how many people went to the movies. (Example 4)

_____

5. **Building on the Essential Question** How can you use the *work backward* problem-solving strategy to solve a two-step equation?

_____

_____

_____

Rate Yourself!

How confident are you about solving equations? Check the box square that applies.

☹ ☹ 😊

☐ ☐ ☐ ☐ ☐

For more help, go online to access a Personal Tutor.

Tutor 💬

Independent Practice

Go online for Step-by-Step Solutions

Solve each equation. Check your solution. (Examples 1–3)

1. $5 = 4a - 7$

 Show your work.

2. $16 = 5x - 9$

3. $3 - 8c = 35$

4. $-\frac{1}{2}x - 7 = -11$

5 $15 - \frac{w}{4} = 28$

6. $-3 - 6x = 9$

7 Larina received a $50 gift card to an online store. She wants to purchase some bracelets that cost $8 each. There will be a $10 overnight delivery fee. Solve $8n + 10 = 50$ to find the number of bracelets she can

purchase. (Example 4) _____

8. LaTasha paid $75 to join a summer golf program. The course where she plays charges $30 per round. Since she is a student, she receives a $10 discount per round. If LaTasha spent $375, use the equation $375 = 20g + 75$ to find how many rounds of golf LaTasha played.

(Example 4) _____

Copy and Solve Solve each equation. Show your work on a separate piece of paper.

9. $\frac{a - 4}{5} = 12$

10. $\frac{n + 3}{8} = -4$

11. $\frac{6 + z}{10} = -2$

12. **MP Reason Abstractly** If Mr. Arenth wants to put new carpeting in the room shown, how many square feet should he order?

_____

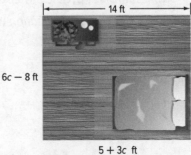

14 ft

$6c - 8$ ft

$5 + 3c$ ft

13. **MP** **Model with Mathematics** Refer to the graphic novel frame below for Exercises a–b.

a. The equation $50 = 28.10 + 0.15m$ represents the additonal number of messages Jacob can send with a budget of $50. Solve the equation to

   find the number of messages he has left to send. _____

Show your work.

b. The equation $50 = 36.50 + 0.10m$ represents the additional number of messages Roberto can send with a budget of $50. Solve the equation

   to find the number of messages he has left to send. _____

🔥 H.O.T. Problems Higher Order Thinking

14. **MP** **Persevere with Problems** Solve $(x + 5)(x + 5) = 49$.
    (*Hint:* There are two solutions.)

    _____

15. **MP** **Model with Mathematics** Write a real-world problem that could be solved by using the equation $3x - 25 = 125$. Then solve the equation.

    _____

    _____

16. **MP** **Use a Counterexample** Determine if the following statement is *true* or *false*. If *false*, provide a counterexample.

    > An equation with an integer coefficient will always
    > have an integer solution.

    _____

    _____

Extra Practice

Solve each equation. Check your solution.

17. $2h + 9 = 21$

$$2h + 9 = 21$$
$$\underline{-9 = -9}$$
$$\frac{2h}{2} = \frac{12}{2}$$
$$h = 6$$

Homework Help ➡

18. $12 - \frac{3}{5}p = -27$

$$12 - \frac{3}{5}p = -27$$
$$\underline{-12 \qquad = -12}$$
$$-\frac{3}{5}p = -39$$
$$\left(-\frac{5}{3}\right)\left(-\frac{3}{5}p\right) = -39\left(-\frac{5}{3}\right)$$
$$p = 65$$

19. $11 = 2b + 17$

20. $-17 = 6p - 5$

21. $2g - 3 = -19$

22. $13 = \frac{g}{3} + 4$

23. $13 - 3d = -8$

24. $-\frac{2}{3}m - 4 = 10$

25. $-5y - 25 = 25$

26. Some friends decide to go to the aquarium together. Each person pays $7.50 to get in. They spend a total of $40 for the shark exhibit. The total cost is $70. Solve $7.5x + 40 = 70$ to find how many people went to the aquarium.

_____

27. **MP Identify Structure** Brent had $26 when he went to the fair. After playing 7 games, he had $15.50 left. Solve $15.50 = 26 - 7p$ to find the price for each game. Then list the Properties of Equality you used to solve the equation.

_____

28. Use the algebra tiles to model the equation $3x + 1 = 7$ on the equation mat below. Then solve the equation.

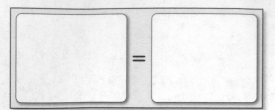

$x = \boxed{}$

29. Determine if the value of the variable is a solution of each equation. Select yes or no.

a. $5x - 4 = 31$, $x = 5.4$ ☐ yes ☐ no

b. $\frac{3}{4}n + 4 = 10$, $n = 8$ ☐ yes ☐ no

c. $-3 + 4y = 7$, $y = 2.5$ ☐ yes ☐ no

Common Core Spiral Review

Solve each equation. Check your solution. 6.EE.7

30. $t - 17 = 5$ **31.** $a - 5 = 14$ **32.** $9 = 5 + x$

Write and solve an equation for each of the following. 6.EE.7

33. Solomon is 9 years younger than his brother. His brother is 21. How old is Solomon? _____

34. Kelly spent $45 more on boots than she did on a pair of jeans. She spent $79.50 on the boots. How much did she spend on the jeans?

_____

35. The product of two integers is 72. If one integer is 18, what is the other integer? _____

Write Two-Step Equations

 Real-World Link

Robotics You want to attend a two-week robotic day camp that costs $700. Your parents will pay the deposit of $400 if you pay the rest in weekly payments of $15. Use the questions below to help you find the number of weeks you will need to make payments.

 Essential Question

WHAT is equivalence?

 Common Core State Standards

Content Standards
8.EE.7, 8.EE.7a, 8.EE.7b

MP Mathematical Practices
1, 2, 3, 4

1. Complete the table below. How much is paid after 2, 3, and 4 weeks?

Payments	Amount Paid
0	$400 + 15(0) = 400$
1	$400 + 15(1) = 415$
2	
3	
4	

2. It will take a long time to solve the problem with a table. Instead, write and solve an equation to find the number of payments p you will need to make.

3. How many payments will you make? ☐

4. Suppose you received $75 in birthday money that you want to use towards the camp. Write and solve an equation to find the number of payments p you will

need to make. _____

Which MP Mathematical Practices did you use?
Shade the circle(s) that applies.

① Persevere with Problems
② Reason Abstractly
③ Construct an Argument
④ Model with Mathematics

⑤ Use Math Tools
⑥ Attend to Precision
⑦ Make Use of Structure
⑧ Use Repeated Reasoning

Translate Sentences into Equations

There are three steps to writing a two-step equation.

Words	Describe the situation. Use only the most important words.
Variable	Define a variable to represent the unknown quantity.
Equation	Translate your verbal model into an algebraic equation.

You know how to write verbal sentences as one-step equations. Some verbal sentences translate into two-step equations.

Examples

Tutor

Translate each sentence into an equation.

1. **Eight less than three times a number is −23.**

Words	Eight less than three times a number is −23.
Variable	Let n represent the number.
Equation	$3n - 8 = -23$

2. **Thirteen is 7 more than one-fifth of a number.**

Words	Thirteen is 7 more than one-fifth of a number.
Variable	Let n represent the number.
Equation	$13 = \frac{1}{5}n + 7$

Show your work.

Got it? **Do these problems to find out.**

a. Fifteen equals three more than six times a number.

b. Ten increased by the quotient of a number and 6 is 5.

c. The difference between 12 and $\frac{2}{3}$ of a number is 18.

STOP and Reflect

Name 3 words that indicate an addition statement.

a. _____

b. _____

c. _____

Examples

3. You buy 3 books that each cost the same amount and a magazine, all for $55.99. You know that the magazine costs $1.99. How much does each book cost?

Words	Three books and a magazine cost $55.99.
Variable	Let b represent the cost of one book.
Equation	$3b + 1.99 = 55.99$

$$3b + 1.99 = 55.99 \quad \text{Write the equation.}$$
$$\underline{-1.99 = -1.99} \quad \text{Subtraction Property of Equality}$$
$$3b = 54.00 \quad \text{Simplify.}$$
$$\frac{3b}{3} = \frac{54.00}{3} \quad \text{Division Property of Equality}$$
$$b = 18 \quad \text{Simplify.}$$

So, the books each cost $18.

4. A personal trainer buys a weight bench for $500 and w weights for $24.99 each. The total cost of the purchase is $849.86. How many weights were purchased?

Words	Bench plus $24.99 per weight equals $849.86
Variable	Let w represent the number of weights.
Equation	$500 + 24.99 \cdot w = 849.86$

$$500 + 24.99w = 849.86 \quad \text{Write the equation.}$$
$$\underline{-500 \qquad\qquad = -500} \quad \text{Subtraction Property of Equality}$$
$$24.99w = 349.86 \quad \text{Simplify.}$$
$$\frac{24.99w}{24.99} = \frac{349.86}{24.99} \quad \text{Division Property of Equality}$$
$$w = 14 \quad \text{Simplify.}$$

So, 14 weights were purchased.

Got it? Do this problem to find out.

Show your work.

d. The current temperature is 54°F. It is expected to rise 2.5°F each hour. In how many hours will the temperature be 84°F?

d. _____

Example

5. Your and your friend's lunch cost $19. Your lunch cost $3 more than your friend's. How much was your friend's lunch?

Words	Your friend's lunch plus your lunch equals $19.
Variable	Let f represent the cost of your friend's lunch.
Equation	$f + f + 3 = 19$

$$f + f + 3 = 19$$ Write the equation.

$$2f + 3 = 19$$ $f + f = 2f$

$$\underline{-3 = -3}$$ Subtraction Property of Equality

$$2f = 16$$ Simplify.

$$\frac{2f}{2} = \frac{16}{2}$$ Division Property of Equality

$$f = 8$$ Simplify.

Your friend spent $8.

Guided Practice

Check ✓

Translate each sentence into an equation. (Examples 1 and 2)

1. One more than three times a number is 7. _____

2. Seven less than one-fourth of a number is −1. _____

3. The quotient of a number and 5, less 10, is 3. _____

4. You already owe $4.32 in overdue rental fees and are returning a movie that is 4 days late. Now you owe $6.48. Define a variable. Then write and solve an equation to find the daily fine for an overdue movie. (Examples 3–5)

_____

_____

5.  **Building on the Essential Question** Why is it important to define a variable before writing an equation?

_____

_____

_____

Independent Practice

Go online for Step-by-Step Solutions

Translate each sentence into an equation. (Examples 1 and 2)

1. Four less than five times a number is equal to 11. _____

2. Fifteen more than half a number is 9. _____

3 Six less than seven times a number is equal to −20. _____

4. Eight more than four times a number is −12. _____

Define a variable. Then write and solve an equation to solve each problem. (Examples 3–5)

5 **Financial Literacy** The cost for a certain music plan is $9.99 per year plus $0.25 per song you download. If you paid $113.74 one year, find the number of songs you downloaded. _____

6. Amy has saved $725 for a new guitar and lessons. Her guitar costs $475, and guitar lessons are $25 per hour. Determine how many hours of lessons she can afford. _____

7. From ground level to the tip of the torch, the Statue of Liberty and its pedestal are 92.99 meters tall. The pedestal is 0.89 meter taller than the statue. How tall is the Statue of Liberty?

_____

8. **MP** **Reason Abstractly** Elsie would like to take snowboarding lessons at Powder Mountain. She has saved $550 for lessons and a junior season pass. How many more semi-private lessons than private lessons can she take? _____

Powder Mountain Ski Resort Snowboarding Lessons	
Semi-Private	$45/lesson
Private	$60/lesson
Junior Season Pass	$315

9. When diving, the peregrine falcon can reach speeds of up to 175 miles per hour. Write and solve equations to find each of the following.

   a. The top speed of a peregrine falcon is 20 miles per hour less than three times the top speed of a cheetah. What is the cheetah's

      top speed? _____

   b. A sailfish can swim up to 1 mile per hour less than one fifth the top speed of a peregrine falcon. Find the top speed that a sailfish

      can swim. _____

   c. The peregrine falcon can reach speeds about 13 miles per hour more than 6 times the speed of the fastest human. What is the

      approximate top speed of the fastest human? _____

H.O.T. Problems Higher Order Thinking

10. **MP Model with Mathematics** If 12 less than 4 times a number is 8, the number is 5. Write a different sentence where the unknown number is

    also 5. _____

11. **MP Persevere with Problems** The ages of three siblings combined is 27. The oldest is twice the age of the youngest. The middle child is 3 years older than the youngest. Write and solve an equation to find the ages of each

    sibling. _____

    _____

12. **MP Model with Mathematics** Write about a real-world situation that can be solved using a two-step equation. Then write the equation and solve the

    problem. _____

    _____

    _____

13. **MP Model with Mathematics** Describe two real-world situations that can be represented by the same two-step equation.

    Situation 1: _____

    _____

    Situation 2: _____

    _____

Extra Practice

Translate each sentence into an equation.

14. Twenty-two less than three times a number is −70. $3n - 22 = -70$ _____

     Words   Twenty-two less than three times a number is −70.

     Variable   Let n represent the number.

     Equation $3n - 22 = -70$

15. The product of a number and 4 increased by 16 is −2. _____

16. Twelve less than the one-fifth of a number is −7. _____

17. Six more than nine times a number is 456. _____

Define a variable. Then write and solve an equation to solve each problem.

18. It costs $13 for admission to an amusement park, plus $1.50 for each ride. If you have a total of $35.50 to spend, what is the greatest number

of rides you can go on? _____

19. Trey went to the batting cages to practice hitting. He rented a helmet for $4 and paid $0.75 for each group of 20 pitches. If he spent a total of $7 at the batting cages, how many groups of pitches did he pay for?

_____

20. **MP** **Make a Conjecture** Hunter and Amado are each trying to save $600 for a summer trip. Hunter started with $150 and earns $7.50 per hour working at a grocery store. Amado has nothing saved, but he earns $12 per hour painting houses.

    a. Make a conjecture about who will take longer to save enough money for

    the trip. Justify your reasoning. _____

    _____

    _____

    b. Write and solve two equations to check your conjecture.

    _____

21. Use the figure to fill in each blank to make a true statement.
The simplified expression for the perimeter of the rectangle is

[_____] .

An equation that can be used to find w is [_____] . The width

of the rectangle is [_____] .

$2w + 3$

w

Perimeter = 36 units

22. Model each situation below with an equation. Select the correct
equation for each situation. Then solve each problem.

a. A company employs 72 workers. It plans to increase the
number of employees by 6 per month until it has twice
its current workforce. How many months will it take to
double the number of employees?

$6m + 72 = 96$	$144 - 72m = 6$
$6m + 72 = 144$	$144 - 6m = 96$

Equation: [_____]   Solution: [_____]

b. Nathan's fish tank holds 144 gallons of water. To clean the tank, he drains
the water at a rate of 6 gallons per minute until the level is two-thirds its
original level. How many minutes does it take to drain the tank for
cleaning?

Equation: [_____]   Solution: [_____]

Common Core Spiral Review

Solve each equation. Check your solution. **7.EE.4**

23. $\dfrac{y}{7} = 22$

24. $\dfrac{a}{6} = -108$

25. $-6 = \dfrac{n}{8} + 1$

26. $-15 = -4p + 9$

27. In a recent NFL game, the Green Bay Packers scored 14 points less than
the Tennessee Titans. Write and solve an equation to find the total points

the Tennessee Titans scored. **6.EE.7** _____

Preseason Week 4	
Team	**Total Points**
Packers	17
Titans	p

Problem-Solving Investigation
Work Backward

CCSS Content Standards
8.EE.7

MP Mathematical Practices
1, 4, 7

Case #1 Game Switcheroo!

Hector and Alex traded video games. Alex gave Hector one fourth of his video games in exchange for 6 video games. Then he sold 3 video games and gave 2 video games to his brother. Alex ended up with 16 video games.

How many video games did Alex have when he started?

Understand *What are the facts?*

- Alex now has 16 games.
- He gave some away, sold some, and traded some.

Plan *What is your strategy to solve this problem?*

Start with the ending number of video games, 16, and work backward.

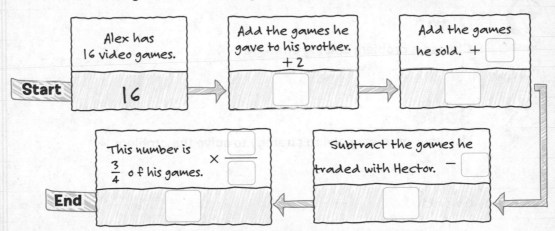

Alex has 16 video games.	Add the games he gave to his brother. + 2	Add the games he sold. + ☐
Start 16	☐	☐

This number is $\frac{3}{4}$ of his games. × ☐	Subtract the games he traded with Hector. − ☐
End ☐	☐

Solve *How can you apply the strategy?*

So, Alex had ☐ video games at the beginning.

Check *Does the answer make sense?*

Start with 20. Perform operations in reverse order.

Analyze the Strategy   Tutor

MP Identify Structure How is working backward similar to solving an equation?

Aurora raised money for a white water rafting trip. Jacy made the first donation. Guillermo's donation was twice Jacy's donation. Rosa's mother tripled what Aurora had raised so far. Now Aurora has $120.

How much did Jacy donate?

Understand

Read the problem. What are you being asked to find?

I need to find _____.

Underline key words and values. What information do you know?

Jacy donated _____. Guillermo _____ and

Rosa's mother _____ the whole amount collected.

Is there any information that you do *not* need to know?

I do not need to know _____.

Plan

Choose a problem-solving strategy.

I will use the _____ strategy.

Solve

Use your problem-solving strategy to solve the problem.

Aurora raised a total of _____.

Go Back Divide that amount by 4. One part is Guillermo's and Jacy's donation and three parts is the amount donated by Rosa's mother. $120 ÷ 4 = _____

Go Back Divide that amount by 3. One part is for Jacy's donation and two parts is the amount that Guillermo donated. _____ ÷ 3 = _____

Jacy was the first to donate. So, Jacy donated _____.

Check

Use information from the problem to check your answer.

Begin with $10 and perform operations in reverse. _____ × 2 = $20;

$20 + $10 = _____; _____ × 3 = $90; _____ + _____ = _____.

Work with a small group to solve the following cases.
Show your work on a separate piece of paper.

Case #3 Financial Literacy

Janelle has $75. She buys jeans that are discounted 30% and then uses an in-store coupon for $10 off the discounted price. After paying $3.26 in sales tax, she receives $17.34 in change.

What was the original price of the jeans?

_____

Case #4 Schedule

Nyoko needs to be at school at 7:45 A.M. It takes her 15 minutes to walk to school, $\frac{5}{12}$ hour to eat breakfast, 0.7 hour to get dressed, and 0.15 hour to shower.

At what time should Nyoko get up to be at school 5 minutes early?

_____

Case #5 Financial Literacy

At the end of the month, Mr. Copley has $1,473.61 in his checking account. His checkbook showed the following transactions.

If he made an initial deposit of $75.00 that is not shown in his checkbook, what was his balance at the beginning of the month?

Chk No.	Date	Payment or Withdrawal		Deposit	
				$150	00
132		$45	79		
		$100	62		
133		$18	48		
				$250	00

Use any strategy!

Case #6 Geometry

Study the pattern below.

Draw the next two figures in the pattern.

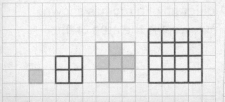

Mid-Chapter Check

Vocabulary Check

1. **MP Be Precise** Define *multiplicative inverse*. Give an example of a number and its multiplicative inverse. (Lesson 1)

_____

_____

2. Fill in the blank in the sentence below with the correct term. (Lesson 2)

   The first step in solving the equation $3x + 4 = 20$ is to _____

   from each side. This is an example of the _____ Property of

   _____.

Skills Check and Problem Solving

Solve each equation. Check your solution. (Lessons 1–2)

3. $\frac{2}{3}x = -8$

   Show your work.

4. $-4.5 = -0.15p$

5. $2\frac{1}{3}c = 2\frac{1}{10}$

6. $3m + 5 = 14$

7. $-2k + 7 = -3$

8. $11 = \frac{1}{3}a + 2$

9. **MP Persevere with Problems** A diagram of a room is shown. If the perimeter of the room is 78 feet, what is the area of the floor of the room? (Lesson 3)

   _____

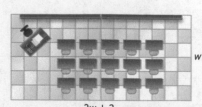

$2w + 3$

w

Inquiry Lab

Equations with Variables on Each Side

Content Standards 8.EE.7, 8.EE.7a

MP Mathematical Practices 1, 3, 5

Inquiry HOW do you use the Properties of Equality when solving an equation using algebra tiles?

Leah bought 4 pens and a bottle of nail color. Her sister bought 2 of the same pens and 4 bottles of nail color, and spent the same amount as Leah. The nail color cost $2. Use algebra tiles to find the cost of each pen.

Hands-On Activity 1

Tools

The equation $4x + 2 = 2x + 8$ represents the real-world situation above. Use algebra tiles to model and solve the equation.

Step 1 Model the equation.

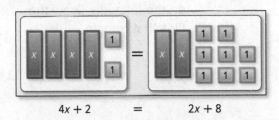

$$4x + 2 \quad = \quad 2x + 8$$

Step 2 Remove ☐ x-tiles from each side of the mat until there are x-tiles on only one side.

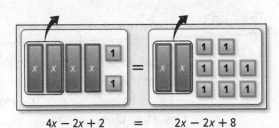

$$4x - 2x + 2 \quad = \quad 2x - 2x + 8$$

Step 3 Remove ☐ 1-tiles from each side of the mat until the x-tiles are by themselves on one side.

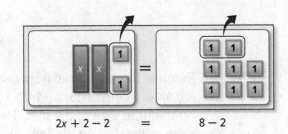

$$2x + 2 - 2 \quad = \quad 8 - 2$$

Step 4 Separate the tiles in ☐ equal groups.

Check $4 \cdot \boxed{} + 2 \overset{?}{=} 2 \cdot \boxed{} + 8$

$$14 = 14 \checkmark$$

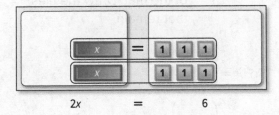

$$2x \quad = \quad 6$$

So, each pen costs $ ☐ .

Hands-On Activity 2

Use algebra tiles to model and solve $3x + 3 = 2x - 3$. Draw the tiles in the blank mats shown. The first step is done for you.

Step 1   Model the equation.

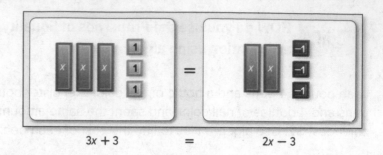

$3x + 3$ $=$ $2x - 3$

Step 2   Remove 2 x-tiles from each side of the mat in Step 1 so that there is an x-tile by itself on the left side. Draw the tiles that remain.

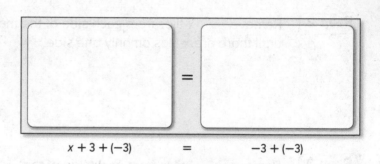

$3x - 2x + 3$ $=$ $2x - 2x - 3$

Step 3   To isolate the x-tile, it is not possible to remove the same number of 1-tiles from each side of the mat. Add three −1-tiles to each side of the mat. Draw the tiles.

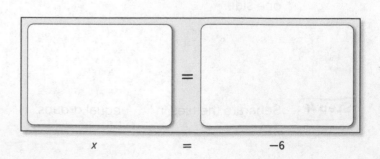

$x + 3 + (-3)$ $=$ $-3 + (-3)$

Step 4   Remove the zero pairs from the left side. There are six −1-tiles on the right side of the mat. The x-tile is isolated on the left side of the mat. Draw the tiles that remain.

x $=$ -6

So, $x =$ ⬚.

Check $3\left(\boxed{} \right) + 3 \overset{?}{=} 2\left(\boxed{} \right) - 3$

$-15 = -15$ ✓   The solution is correct.

Investigate

Collaborate

MP **Use Math Tools** Work with a partner. Model and solve each equation. Show your work using drawings. Write the solution below the mat.

1. $x + 2 = 2x + 1$

2. $2x + 7 = 3x + 4$

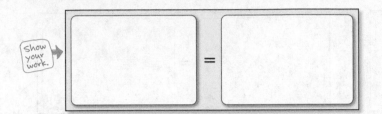

$x = $ _____

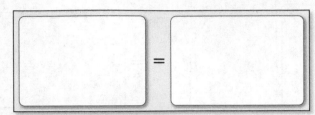

$x = $ _____

3. $2x - 5 = x - 7$

4. $x + 6 = 3x - 2$

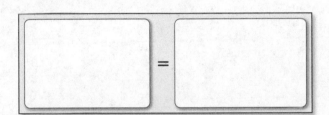

$x = $ _____

$x = $ _____

5. $8 + x = 3x$

6. $3x + 6 = 6x$

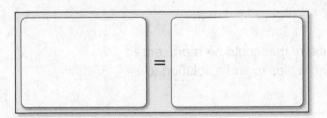

$x = $ _____

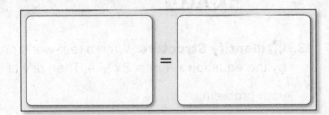

$x = $ _____

7. $3x + 3 = x - 5$

8. $2x + 5 = 4x - 1$

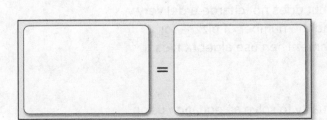

$x = $ _____

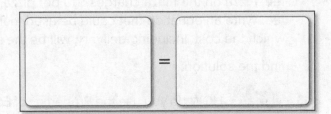

$x = $ _____

Inquiry Lab Equations with Variables on Each Side   **143**

Analyze and Reflect

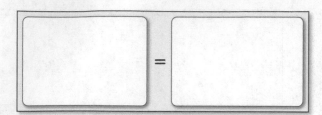

Work with a partner. One of you should solve the following equations by removing 1-tiles first. The other one should solve the equations by removing x-tiles first. Compare your answers.

9. $x + 4 = 3x - 4$

10. $4x + 2 = x - 4$

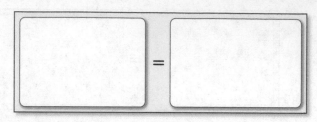

$x =$ _____

$x =$ _____

11. **MP Reason Inductively** Does it matter whether you remove x-tiles or 1-tiles first? Is one way more convenient? Explain. _____

_____

_____

_____

12. **MP Use Math Tools** Explain why you can remove an x-tile from each side of

the mat. _____

Create

13. **MP Identify Structure** Write a real-world problem that could be represented by the equation $x + 4 = 3x - 4$. Then use algebra tiles to find a solution to

your problem. _____

_____

_____

14. **MP Identify Structure** Pizza Shack charges $8 per pizza with a $4 delivery fee. Pizza on the Plaza charges $10 per pizza, but does not charge a delivery fee. Write an equation that could be used to find the number of pizzas for which the cost, including delivery, will be the same. Then use algebra tiles to

find the solution. _____

15. **Inquiry** HOW do you use the Properties of Equality to solve an equation using

algebra tiles? _____

_____

Solve Equations with Variables on Each Side

Cell Phones A wireless company offers two cell phone plans. Plan A charges $24.95 per month plus $0.10 per minute for calls. Plan B charges $19.95 per month plus $0.20 per minute. Use the questions to find when the two plans cost the same.

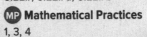

1. Complete the table.

Minutes (m)	Plan A $24.95 + 0.10m$	Plan B $19.95 + 0.20m$
10		
20		
30		
40		
50		
60		
70		

2. For what value(s) does Plan A cost less?

_____

3. For what value(s) does Plan B cost less?

_____

4. For what value(s) do both Plans cost the same?

_____

Which MP **Mathematical Practices** did you use?
Shade the circle(s) that applies.

① Persevere with Problems   ⑤ Use Math Tools

② Reason Abstractly   ⑥ Attend to Precision

③ Construct an Argument   ⑦ Make Use of Structure

④ Model with Mathematics   ⑧ Use Repeated Reasoning

Equations with Variables on Each Side

Some equations, like $8 + 4d = 5d$, have variables on each side of the equals sign. To solve, use the properties of equality to write an equivalent equation with the variables on one side of the equals sign. Then solve the equation.

Examples

Tutor

1. **Solve $8 + 4d = 5d$. Check your solution.**

$$8 + 4d = 5d \qquad \text{Write the equation.}$$
$$\underline{-4d = -4d} \qquad \text{Subtraction Property of Equality}$$
$$8 = d \qquad \text{Simplify by combining like terms.}$$

| Subtract $4d$ from the left side of the equation to isolate the variable. | Subtract $4d$ from the right side of the equation to keep it balanced. |

To check your solution, replace d with 8 in the original equation.

$$\text{Check } \quad 8 + 4d = 5d \qquad \text{Write the original equation.}$$
$$8 + 4(8) \overset{?}{=} 5(8) \qquad \text{Replace } d \text{ with 8.}$$
$$40 = 40 \checkmark \qquad \text{The sentence is true.}$$

2. **Solve $6n - 1 = 4n - 5$.**

$$6n - 1 = 4n - 5 \qquad \text{Write the equation.}$$
$$\underline{-4n \quad = -4n} \qquad \text{Subtraction Property of Equality}$$
$$2n - 1 = -5 \qquad \text{Simplify.}$$
$$\underline{+1 = +1} \qquad \text{Addition Property of Equality}$$
$$2n = -4 \qquad \text{Simplify.}$$
$$n = -2 \qquad \text{Mentally divide each side by 2.}$$

$$\text{Check } \quad 6n - 1 = 4n - 5 \qquad \text{Write the original equation.}$$
$$6(-2) - 1 \overset{?}{=} 4(-2) - 5 \qquad \text{Replace } n \text{ with } -2.$$
$$-13 = -13 \checkmark \qquad \text{The sentence is true.}$$

Show your work.

Got it? **Do these problems to find out.**

Solve each equation. Check your solution.

    a. $8a = 5a + 21$ **b.** $3x - 7 = 8x + 23$

a. _____

b. _____

Example

Tutor

3. Green's Gym charges a one time fee of $50 plus $30 per session for a personal trainer. A new fitness center charges a yearly fee of $250 plus $10 for each session with a trainer. For how many sessions is the cost of the two plans the same?

Words	fee of $50 plus $30 per session	is the same as	a fee of $250 plus $10 per session.

Variable   Let s represent the number of sessions.

Equation $50 + 30s = 250 + 10s$

$50 + 30s = 250 + 10s$ Write the equation.

$\underline{ -10s = -10s}$ Subtraction Property of Equality

$50 + 20s = 250$ Simplify.

$\underline{-50 = -50}$ Addition Property of Equality

$20s = 200$ Simplify.

$\dfrac{20s}{20} = \dfrac{200}{20}$ Division Property of Equality

$s = 10$ Simplify.

So, the cost is the same for 10 personal trainer sessions.

Check

Green's Gym: $50 plus 10 sessions at $30 per session

$50 + 10 \cdot 30 = 50 + 300$

$= \$350$

new fitness center: $250 plus 10 sessions at $10 per session

$250 + 10 \cdot 10 = 250 + 100$

$= \$350 ✓$

Got it? Do this problem to find out.

c. The length of a flag is 0.3 foot less than twice its width. If the perimeter is 14.4 feet longer than the width, find the dimensions of the flag.

Show your work.

c. _____

Equations with Rational Coefficients

In some equations, the coefficients of the variables are rational numbers. Remember when working with fractions, you need to have a common denominator before you add or subtract.

Example

4. Solve $\frac{2}{3}x - 1 = 9 - \frac{1}{6}x$.

$\frac{4}{6}x - 1 = 9 - \frac{1}{6}x$ The common denominator of the coefficients is 6. Rewrite the equation.

$+\frac{1}{6}x = +\frac{1}{6}x$ Addition Property of Equality

$\frac{5}{6}x - 1 = 9$ Simplify.

$+1 = +1$ Addition Property of Equality

$\frac{5}{6}x = 10$ Simplify.

$\left(\frac{6}{5}\right)\frac{5}{6}x = 10\left(\frac{6}{5}\right)$ Multiplication Property of Equality

$x = 12$ Simplify.

 Show your work.

e. _____

f. _____

Got it? Do these problems to find out.

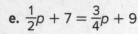

e. $\frac{1}{2}p + 7 = \frac{3}{4}p + 9$

f. $-\frac{5}{4}c - \frac{1}{2} = -\frac{3}{4} + \frac{5}{8}c$

Guided Practice

 Check ✓

Solve each equation. Check your solution. (Examples 1, 2, 4)

1. $5n + 9 = 2n$

2. $7y - 8 = 6y + 1$

3. $\frac{3}{5}x - 15 = \frac{6}{5}x + 12$

 Show your work.

4. EZ Car Rental charges $40 a day plus $0.25 per mile. Ace Rent-A-Car charges $25 a day plus $0.45 per mile. What number of miles results in the same cost for one day? (Example 3) _____

5. **Building on the Essential Question** How is solving an equation with the variable on each side similar to solving a two-step equation? _____

_____

_____

Rate Yourself!

How well do you understand how to solve equations? Circle the figure that applies.

Clear   Somewhat Clear   Not So Clear

For more help, go online to access a Personal Tutor.

Independent Practice

Go online for Step-by-Step Solutions

eHelp

Solve each equation. Check your solution. (Examples 1, 2, 4)

1. $7a + 10 = 2a$

Show your work.

2. $11x = 24 + 8x$

3. $8y - 3 = 6y + 17$

4. $5p + 2 = 4p - 1$

5. $15 - \frac{1}{6}n = \frac{1}{6}n - 1$

6. $3 - \frac{2}{9}b = \frac{1}{3}b - 7$

7. Nine fewer than half a number is five more than four times the number.
Define a variable, write an equation, and solve to find the number. (Example 3)

_____

8. The table shows ticket prices for the local minor
league baseball team for fan club members and
non-members. For how many tickets is the cost the
same for club members and non-members? (Example 3)

_____

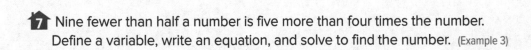

	Ticket Prices	
	Club Members	**Non-Club Members**
Membership Fee (one-time)	$30	none
Ticket Price	$3	$6

9. **MP Multiple Representations** Refer to the square at the right.

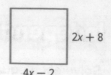

2x + 8

4x − 2

   a. **Words** Explain a method you could use to find the value of x.

_____

_____

   b. **Symbols** Write an equation to find the side length of the square.

_____

   c. **Algebra** What is the side length of the square?

_____

 ## H.O.T. Problems Higher Order Thinking

10. **MP Find the Error** Alma is solving the equation $4a − 5 = 2a − 3$.
    Circle her mistake and correct it.

_____

_____

_____

_____

$$4a − 5 = 2a − 3$$
$$4a − 2a − 5 = 3$$
$$2a − 5 = 3$$
$$2a = 8$$

11. **MP Model with Mathematics** Write a real-world problem that can be
    solved using the equation $5x = 3x + 20$. _____

_____

_____

12. **MP Persevere with Problems** Find the area of the rectangle

    at the right. _____

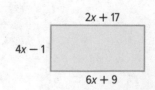

2x + 17

4x − 1

6x + 9

13. **MP Model with Mathematics** Write two equations so that each have

    variables on both sides and a solution of $\frac{1}{2}$. _____

_____

Extra Practice

Solve each equation. Check your solution.

14. $9g - 14 = 2g$

Homework Help →

$$9g - 14 = 2g$$
$$\underline{-9g \qquad = -9g}$$
$$\frac{-14}{-7} = \frac{-7g}{-7}$$
$$2 = g$$

15. $-6f + 13 = 2f - 11$

16. $2.5h - 15 = 4h$

17. $2z - 31 = -9z + 24$

18. Will averages 18 points a game and is the all-time scoring leader on his team with 483 points. Tom averages 21 points a game and is currently second on the all-time scorers list with 462 points. If both players continue to play at the same rate, how many more games will it take until Tom and Will have scored the same number of total points?

19. Eighteen less than three times a number is twice the number. Define a variable, write an equation, and solve to find the number.

MP Reason Abstractly Write an equation to find the value of *x* so that each pair of polygons has the same perimeter. Then solve.

20.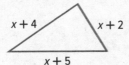

21.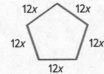

22. The two regular polygons below have the same perimeter.

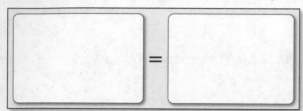

Use the algebra tiles to model an equation on the equation mat below that can be used to find x. Then solve the equation.

$x = \boxed{}$

23. Company A charges $28.50 plus $18 a room to clean carpet. Company B charges $16.50 plus $20 a room. Determine if each statement is true or false.

a. For 4 rooms, carpet cleaner B is cheaper. ☐ True ☐ False

b. For 5 rooms, carpet cleaner A is cheaper. ☐ True ☐ False

c. The equation $28.5 + 18x = 16.5 + 20x$ can be solved to find the number of rooms for which the total cost is the same. ☐ True ☐ False

d. For 6 rooms, both carpet cleaners charge the same amount. ☐ True ☐ False

CCSS Common Core Spiral Review

Use the Distributive Property to write each expression as an equivalent expression. 7.EE.1

24. $6(x + 5) = $ _____

25. $-8(y - 1) = $ _____

26. $-3(-5z + 12) = $ _____

27. $\frac{1}{3}(6z + 10) = $ _____

Solve Multi-Step Equations

Lacrosse Coach Everly wants to order uniform shirts for all the players *p* on her women's lacrosse team. Each shirt costs $20. There is an additional cost *d* for a player to put her name on the shirt. Use the steps below to write an equation for the total cost *c* if every player on the team orders a shirt with her name on it.

1. Circle the variables above and underline what they represent.

2. Write an expression that represents the cost of one shirt with a player's name on it.

$$\boxed{} + \boxed{}$$

cost of shirt + cost of name

3. Use the expression to write an equation that can be used to find the total cost if every player on the team orders a shirt with her name on it.

$$\boxed{}\left(\boxed{} + \boxed{}\right) = c$$

number of players (cost of shirt + cost of name) = total cost

4. Suppose the total cost for 15 players to buy shirts is $420. Write an equation to show the total cost of the shirts if all of the players put their names on the shirts.

_____

_____

 Essential Question

WHAT is equaivalence?

 Vocabulary

null set
identity

Math Symbols
Ø null set
{ } empty set

Common Core State Standards

Content Standards
8.EE.7, 8.EE.7a, 8.EE.7b

MP Mathematical Practices
1, 2, 3, 4

where did it go?

Which MP Mathematical Practices did you use?
Shade the circle(s) that applies.

① Persevere with Problems
② Reason Abstractly
③ Construct an Argument
④ Model with Mathematics
⑤ Use Math Tools
⑥ Attend to Precision
⑦ Make Use of Structure
⑧ Use Repeated Reasoning

Solve Multi-Step Equations

Some equations contain expressions with grouping symbols. To solve these equations, first expand the expression using the Distributive Property. Then collect like terms if needed, and solve the equation using the Properties of Equality.

Example

Tutor

1. Solve $15(20 + d) = 420$.

$15(20 + d) =$	420	Write the equation.
$300 + 15d =$	420	Distributive Property
$-\,300$	$=-\,300$	Subtraction Property of Equality
$15d =$	120	Simplify.
$\dfrac{15d}{15} =$	$\dfrac{120}{15}$	Division Property of Equality
$d =$	8	Simplify.

Show your work.

Got it? Do these problems to find out.

a. $-3(9 + x) = 33$ b. $5(a - 7) = 25$

a. _____

b. _____

Number of Solutions

	Null Set	One Solution	Identity
Words	no solution	one solution	infinitely many solutions
Symbols	$a = b$	$x = a$	$a = a$
Example	$3x + 4 = 3x$ $4 = 0$ Since $4 \neq 0$, there is no solution.	$2x = 20$ $x = 10$	$4x + 2 = 4x + 2$ $2 = 2$ Since $2 = 2$, the solution is all numbers.

Some equations have no solution. When this occurs, the solution is the **null set** or empty set and is shown by the symbol ∅ or { }. Other equations may have every number as their solution. An equation that is true for every value of the variable is called an **identity**.

Examples

2. Solve $6(x - 3) + 10 = 2(3x - 4)$.

$6(x - 3) + 10 = 2(3x - 4)$	Write the equation.
$6x - 18 + 10 = 6x - 8$	Distributive Property
$6x - 8 = 6x - 8$	Collect like terms.
$\underline{+8 = \quad +8}$	Addition Property of Equality
$6x = 6x$	Simplify.
$\dfrac{6x}{6} = \dfrac{6x}{6}$	Division Property of Equality
$x = x$	Simplify.

The statement $x = x$ is *always* true. The equation is an identity and the solution set is all numbers.

Check	$6(x - 3) + 10 = 2(3x - 4)$	Write the original equation.
	$6(5 - 3) + 10 \overset{?}{=} 2[3(5) - 4]$	Substitute any value for x.
	$6(2) + 10 \overset{?}{=} 2(15 - 4)$	Simplify.
	$22 = 22$ ✓	

3. Solve $8(4 - 2x) = 4(3 - 5x) + 4x$.

$8(4 - 2x) = 4(3 - 5x) + 4x$	Write the equation.
$32 - 16x = 12 - 20x + 4x$	Distributive Property
$32 - 16x = 12 - 16x$	Collect like terms.
$\underline{+16x = \quad +16x}$	Addition Property of Equality
$32 = 12$	Simplify.

The statement $32 = 12$ is *never* true. The equation has no solution and the solution set is ∅.

Check	$8(4 - 2x) = 4(3 - 5x) + 4x$	Write the equation.
	$8[4 - 2(2)] \overset{?}{=} 4[3 - 5(2)] + 4(2)$	Substitute any value for x.
	$8(0) \overset{?}{=} 4(-7) + 8$	Simplify.
	$0 \neq -20$ ✓	Since $0 \neq -20$, the equation has no solution.

Got it? Do these problems to find out.

c. $3(6 - 4x) = -2(6x - 9)$ **d.** $2(3x + 5) = 5(2x - 4) - 4x$

Copyright © McGraw-Hill Education

STOP and Reflect

How do you know if the solution $5 = 0$ indicates no solution, one solution, or infinitely many solutions?

Show your work.

c. _____

d. _____

Example

4. At the fair, Hunter bought 3 snacks and 10 ride tickets. Each ride ticket costs $1.50 less than a snack. If he spent a total of $24.00, what was the cost of each snack?

Write an equation to represent the problem.

$3s + 10(s - 1.5) = 24$	Write the equation.
$3s + 10s - 15 = 24$	Distributive Property
$13s - 15 = 24$	Collect like terms.
$\underline{+15 = +15}$	Addition Property of Equality
$13s = 39$	Simplify.
$\dfrac{13s}{13} = \dfrac{39}{13}$	Division Property of Equality
$s = 3$	Simplify.

So, the cost of each snack was $3.

Guided Practice

Check ✓

Solve each equation. Check your solution. (Examples 1–3)

1. $-8(w - 6) = 32$

Show your work.

2. $8z - 22 = 3(3z + 11) - z$

3. Mr. Richards' class is holding a canned food drive for charity. Juliet collected 10 more cans than Rosana. Santiago collected twice as many cans as Juliet. If they collected 130 cans altogether, how many cans did

Juliet collect? (Example 4) _____

4. ⓔ **Building on the Essential Question** How many possible solutions are there to a linear equation in one variable? Describe each one.

_____

_____

_____

_____

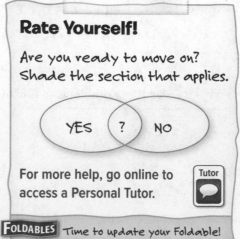

Rate Yourself!

Are you ready to move on? Shade the section that applies.

YES ? NO

For more help, go online to access a Personal Tutor. Tutor 💬

FOLDABLES Time to update your Foldable!

Independent Practice

Go online for Step-by-Step Solutions

Solve each equation. Check your solution. (Examples 1–3)

1. $-12(k + 4) = 60$

 Show your work.

2. $8(3a + 6) = 9(2a - 4)$

3. $\frac{1}{3}h - 4\left(\frac{2}{3}h - 3\right) = \frac{2}{3}h - 6$

4. $8(c - 9) = 6(2c - 12) - 4c$

Copy and Solve Solve each equation. Show your work on a separate piece of paper. (Examples 2 and 3)

5. $-10y + 18 = -3(5y - 7) + 5y$

6. $8(t + 2) - 3(t - 4) = 6(t - 7) + 8$

7. $4(5 + 2x) - 5 = 3(3x + 7)$

8. $6(2x - 8) + 3 = 15$

9. The school has budgeted $2,000 for an end-of-year party at the local park. The cost to rent the park shelter is $150. How much can the student council spend per student on food if each of the 225 students receives a $3.50 gift? (Example 4) _____

10. **MP Reason Abstractly** The table shows the number of students in each homeroom.

a. Write an equation to find the number of students in Mr. Boggs' homeroom if the total number of students is 90. _____

_____

b. Solve the equation from part **a** to find the number of students in Mr. Boggs' homeroom. _____

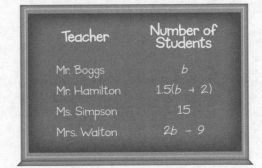

Teacher	Number of Students
Mr. Boggs	b
Mr. Hamilton	$1.5(b + 2)$
Ms. Simpson	15
Mrs. Walton	$2b - 9$

11. **MP Model with Mathematics** Refer to the graphic novel frame below for Exercises a–b.

I've already sent 65 texts this month.

I've sent 54.

My plan is $20/month and $0.15/text. Roberto's plan is $30/month and $0.10/text.

a. Write an equation that can be used to determine the number of text messages Jacob and Roberto can send for their plans to cost the same.

_____

b. Solve the equation from part **a** to find the number of text messages each person can send for their costs to be the same.

_____

H.O.T. Problems Higher Order Thinking

12. **MP Reason Inductively** Does a multi-step equation *always*, *sometimes*, or *never* have a solution? Explain your reasoning.

_____

_____

13. **MP Persevere with Problems** The perimeter of a rectangle is $8(2x + 1)$ inches. The length of the sides of the rectangle are $3x + 4$ inches and $4x + 3$ inches. Write and solve an equation to find the length of each side of the rectangle.

_____

_____

14. **MP Model with Mathematics** Write a real-world problem that can be solved using the Distributive Property. Then write and solve an equation for your real-world situation. _____

_____

_____

Extra Practice

Solve each equation. Check your solution.

15. $9(j - 4) = 81$

$$9j - 36 = 81$$
$$+36 = +36$$
$$9j = 117$$
$$\frac{9j}{9} = \frac{117}{9}$$
$$j = 13$$

16. $8(4q - 5) - 7q = 5(5q - 8)$

$$32q - 40 - 7q = 25q - 40$$
$$25q - 40 = 25q - 40$$
$$-25q = -25q$$
$$-40 = -40$$

The solution set is all numbers.

17. $\frac{1}{2}r + 2\left(\frac{3}{4}r - 1\right) = \frac{1}{4}r + 6$

18. $-5(3m + 6) = -3(4m - 2)$

19. $-7(k + 9) = 9(k - 5) - 14k$

20. $10p - 2(3p - 6) = 4(3p - 6) - 8p$

21. $12(x + 3) = 4(2x + 9) + 4x$

22. $0.2(x + 50) - 6 = 0.4(3x + 20)$

23. **MP Identify Structure** Give an example of a multi-step equation for each of the following solutions.

a. all numbers _____

b. null set _____

24. The table shows expressions to represent the number of students involved in different activities. The number of students involved in sports and student council is equal to the number of students involved in band and drama club. Model the situation with an equation. Select the correct expression for each box. Then solve the equation.

Activity	Number of Students
band	$3n - 2$
drama club	$2(2n + 1)$
sports	$5n + 7$
student council	n

[_____] + [_____] = [_____] + [_____]

$n =$ [_____]

How many students participate in each activity?

sports: [_____]   band: [_____]

student council: [_____]   drama club: [_____]

25. The figures below have the same perimeter. Determine if each statement is true or false.

a. The value of x is 2.   ☐ True ☐ False

b. The perimeter of the triangle is 36 units.   ☐ True ☐ False

c. The perimeter of the square is 32 units.   ☐ True ☐ False

(triangle labeled $x + 6$, $5x$, $3x + 3$; square labeled $x + 6$)

Solve each inequality. Graph the solution set on a number line. 7.EE.4b

26. $a - 5 < 2$ _____

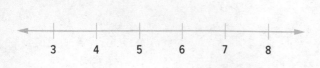

27. $x - 9 \geq -12$ _____

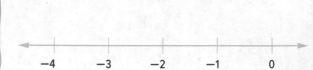

28. $5y \leq -30$ _____

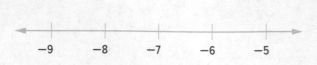

29. $-\dfrac{n}{4} > -2$ _____

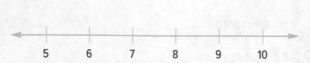

21ST CENTURY CAREER
in Design

Skateboard Designer

If you love the sport of skateboarding, and you are creative and have strong math skills, you should think about a career designing skateboards. A skateboard designer applies engineering principles and artistic ability to design high-performance skateboards that are both strong and safe. To have a career in skateboard design, you should study physics and mathematics and have a good understanding of skateboarding.

College & Career
READINESS

Is This the Career for You?

Are you interested in becoming a skateboard designer? Take some of the following courses in high school.

◆ Digital Design
◆ Geometry
◆ Physics
◆ Trigonometry

Turn the page to find out how math relates to a career in Design.

MP It's Great to Skate

Use the information in the table to solve each problem.

1. The total width of two standard shortboards and a technical shortboard is 23.5 inches. Write an equation to represent the situation.

_____

2. Solve the equation from Exercise 1 to find the width of a standard shortboard.

_____

3. The total length of two longboards and a standard shortboard is 113.4 inches. Write and solve an equation to find the length of a longboard. _____

_____

4. The total width of three technical shortboards is 4.5 inches more than the total width of two longboards. Write and solve an equation to find the width of a longboard.

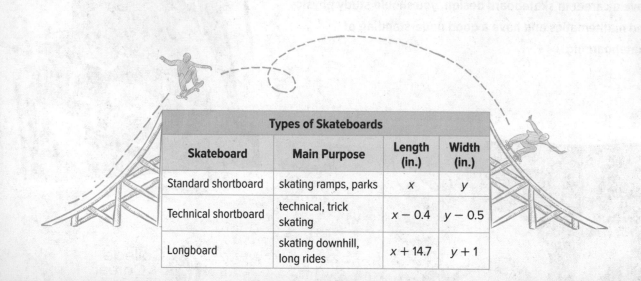

Types of Skateboards			
Skateboard	Main Purpose	Length (in.)	Width (in.)
Standard shortboard	skating ramps, parks	x	y
Technical shortboard	technical, trick skating	$x - 0.4$	$y - 0.5$
Longboard	skating downhill, long rides	$x + 14.7$	$y + 1$

MP Career Project

It's time to update your career portfolio! Describe the skills that would be necessary for a skateboard designer to possess. Determine whether this type of career would be a good fit for you.

_____

_____

_____

_____

_____

What problem-solving skill might you use as a skateboard designer?

_____

_____

_____

_____

Chapter Review

Vocabulary Check

Unscramble each of the clue words. After unscrambling all of the terms, use the numbered letters to find the name of a famous mathematician.

OCIFCTIENEF
⬜⬜⬜⬜⬜⬜⬜⬜⬜⬜⬜
　　　　　　2　　　　　11

TECVALMIITUPLI SEVNEIR
⬜⬜⬜⬜⬜⬜⬜⬜⬜⬜⬜⬜⬜⬜
　　　　　　　　　　6

⬜⬜⬜⬜⬜⬜⬜
　13　　　3

DEYTINTI
⬜⬜⬜⬜⬜⬜⬜
4　　　9

LLUNETS
⬜⬜⬜　⬜⬜⬜

PERRITPEOS
⬜⬜⬜⬜⬜⬜⬜⬜⬜⬜
　　　　　　　　　1

REABIASVL
⬜⬜⬜⬜⬜⬜⬜⬜
　　　　　　10　5

PEOCILRACR
⬜⬜⬜⬜⬜⬜⬜⬜⬜⬜
　　　　　　12　8　7

⬜⬜⬜　⬜⬜⬜⬜⬜　⬜⬜ W ⬜⬜⬜
1　2　3　　4　5　6　7　8　　9　10　11　12　13

Complete each sentence using one of the unscrambled words above.

1. The _____ is the numerical factor of a term that contains a variable.

2. Another name for the reciprocal is the _____.

3. In mathematics, statments that are true for all numbers are _____.

4. When writing an equation from a real-world problem, it is important to define the _____.

Use Your FOLDABLES

Use your Foldable to help review the chapter.

Tape here

Solving Equations

Tab 1

Write About It

Tab 2

Solve

$6(x - 3) + 10 = 2(4x - 5)$

Got it?

Number the steps in the order needed to solve each equation. Then solve the equation.

1. $5(x + 3) = 170$

2 Subtract 15 from each side.

1 Multiply x and 3 by 5.

3 Divide each side by 5.

$x = \underline{31}$

 The first one is done for you.

2. $2p - 9 = 6p + 7$

___ Divide each side by 4.

___ Subtract 7 from each side.

___ Subtract $2p$ from each side.

$p = \underline{\hspace{1cm}}$

3. $-\frac{2}{3}(a + 3) = \frac{5}{3}a - 19$

___ Add $\frac{2}{3}a$ to each side.

___ Add 19 to each side.

___ Multiply a and 3 by $-\frac{2}{3}$.

___ Multiply each side by $\frac{3}{7}$.

$a = \underline{\hspace{1cm}}$

Power Up! Performance Task

Bowling Alley Fun

The Bradley family goes bowling once per month. They each bowl the same number of games.

Write your answers on another piece of paper. Show all of your work to receive full credit.

Part A
Five of the family members will bowl this month. The prices at Dan's Alley are listed in the table. The total budget for the outing is $115. The budget for gas and food is $25. Write an equation to find how many games each family member can bowl. Let g represent the number of games bowled by each family member. Then solve your equation.

Dan's Alley	
Cost per game	$5
Shoe rental (one-time fee per visit)	$3

Part B
The next month, the family tries the Bowling Palace, which is on the other side of town. They set aside $30 for gas and food. Only four family members are bowling this time. The total budget for this outing is $106. The prices at Bowling Palace are listed in the table. Write an equation to find how many games each family member can bowl. Let g represent the number of games bowled by each family member. Then solve your equation.

Bowling Palace	
Cost per game	$2
Shoe rental (one-time fee per visit)	$9

Part C
The following month, Mr. Bradley goes bowling by himself. He debates on which bowling alley to select based only on what it will cost to bowl at each alley. How many games would he need to bowl to pay an equal amount at each bowling alley?

Part D
Given the equation $6(8g + 2) + 45 = 153$, write a scenario similar to Parts A and B that could be represented by the equation. Solve for g and explain what each number in the equation represents, including the solution.

Reflect

Use what you learned about equivalence to complete the graphic organizer.
Draw or write an example for each category.

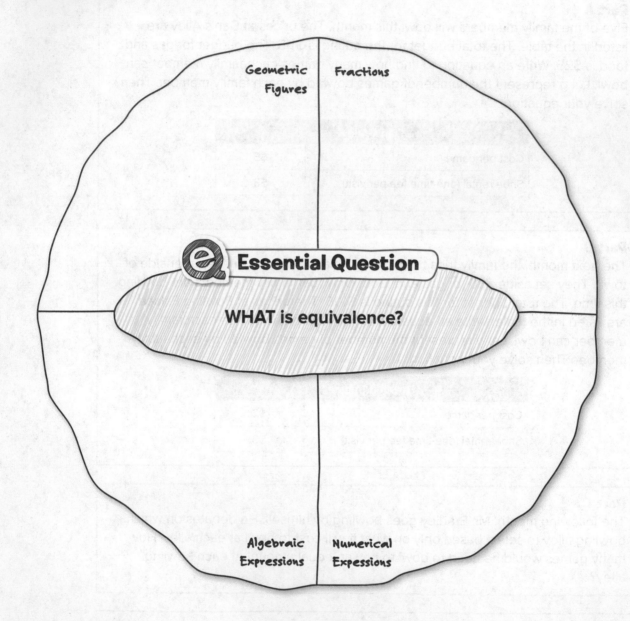

Geometric Figures

Fractions

Essential Question

WHAT is equivalence?

Algebraic Expressions

Numerical Expressions

 Answer the Essential Question. WHAT is equivalence?

Chapter 3
Equations in Two Variables

 Essential Question

WHY are graphs helpful?

 Common Core State Standards

Content Standards
8.EE.5, 8.EE.6, 8.EE.8, 8.EE.8a, 8.EE.8b, 8.EE.8c, 8.F.2, 8.F.3, 8.F.4, 8.F.5

MP **Mathematical Practices**
1, 2, 3, 4, 5, 7

 Math in the Real World

Games Bobbie is playing Attack of the Seas with Monique. Bobbie placed his ship between A5 and C5. To earn points, Monique guesses a point to locate one of Bobbie's ships.

Draw Bobbie's ship on the coordinate plane. If Monique guesses C8, will she locate one of Bobbie's ships? _____

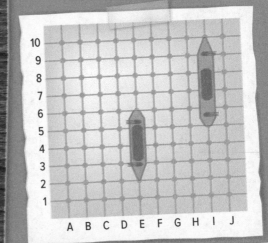

FOLDABLES **Study Organizer**

 Cut out the Foldable on page FL7 of this book.

 Place your Foldable on page 256.

 Use the Foldable throughout this chapter to help you learn about equations in two variables.

Vocabulary

constant of proportionality	point-slope form	standard form
constant of variation	rise	substitution
constant rate of change	run	systems of equations
direct variation	slope	x-intercept
linear relationships	slope-intercept form	y-intercept

Review Vocabulary

Rates A rate is a ratio that compares two quantities with different kinds of units. Some common rates are $\frac{miles}{hour}$, $\frac{price}{ounce}$, and $\frac{meters}{second}$.

Unit Rates A rate is a unit rate when it has a denominator of 1 unit. You can find the unit rate by writing a ratio, then dividing the numerator by the denominator.

Circle the correct answer for each unit rate.

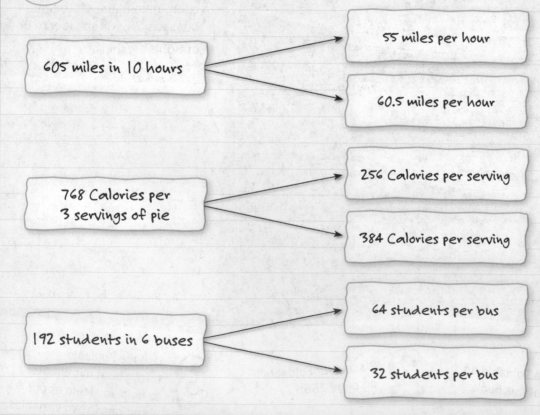

605 miles in 10 hours → 55 miles per hour / 60.5 miles per hour

768 Calories per 3 servings of pie → 256 Calories per serving / 384 Calories per serving

192 students in 6 buses → 64 students per bus / 32 students per bus

What Do You Already Know?

List three things you already know about equations in two variables in the first section. Then list three things you would like to learn about equations in two variables in the second section.

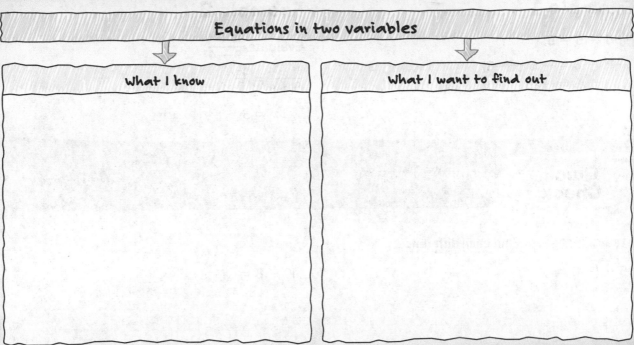

Equations in two variables

What I know	What I want to find out

When Will You Use This?

Here are a few examples of how systems of equations are used in the real world.

Activity 1 Does your state have a state fair? Use the Internet to find the cost of ride tickets or a ride wrist band for a state fair. Which one do you think is a better deal?

_____

_____

_____

Activity 2 Go online at **connectED.mcgraw-hill.com** to read the graphic novel **A Fair Deal**. What is the price of a ride wristband?

_____

Hana and Enrique in

A Fair Deal

WOW! I want to go on THAT!

Try the Quick Check below.
Or, take the Online Readiness Quiz.

 Check ✓

Common Core Review 7.NS.1

Example 1

Find −15 − 8.

$-15 - 8 = -15 + (-8)$ To subtract 8, add −8.

$\qquad = -23$ Simplify.

Example 2

Evaluate $\dfrac{11 + 4}{9 - 4}$.

$\dfrac{11 + 4}{9 - 4} = \dfrac{15}{5}$ Simplify the numerator and denominator.

$\qquad = 3$ Simplify.

Quick Check

Subtract Integers   Find each difference.

1. $5 - (-4) =$ _____

2. $10 - 8 =$ _____

 Show your work.

3. $-4 - 3 =$ _____

4. $-6 - (-2) =$ _____

5. $12 - 6 =$ _____

6. $-5 - (-3) =$ _____

Numerical Expressions   Evaluate each expression.

7. $\dfrac{6 - 2}{5 + 5} =$ _____

8. $\dfrac{7 - 4}{8 - 4} =$ _____

9. $\dfrac{3 - 1}{1 + 9} =$ _____

10. $\dfrac{5 + 7}{8 - 6} =$ _____

11. $\dfrac{2 - 4}{3 + 2} =$ _____

12. $\dfrac{1 - 5}{8 - 2} =$ _____

How Did You Do?

Which problems did you answer correctly in the Quick Check? Shade those exercise numbers below.

(1) (2) (3) (4) (5) (6) (7) (8) (9) (10) (11) (12)

Constant Rate of Change

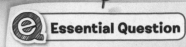

Real-World Link

Music Marcus can download two songs from the Internet each minute. This is shown in the table below.

Time (minutes), x	0	1	2	3	4
Number of Songs, y	0	2	4	6	8

1. Compare the change in the number of songs y to the change in time x. What is the rate of change?

_____

_____

2. Graph the ordered pairs from the table on the graph shown. Label the axes. Then describe the pattern shown on the graph.

_____

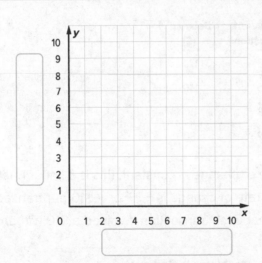

Essential Question

WHY are graphs helpful?

Vocab

Vocabulary

linear relationship
constant rate of change

CCSS **Common Core State Standards**

Content Standards
Preparation for 8.EE.5

MP **Mathematical Practices**
1, 3, 4, 5

Which **MP** **Mathematical Practices** did you use?
Shade the circle(s) that applies.

① Persevere with Problems

② Reason Abstractly

③ Construct an Argument

④ Model with Mathematics

⑤ Use Math Tools

⑥ Attend to Precision

⑦ Make Use of Structure

⑧ Use Repeated Reasoning

Linear Relationships

Relationships that have straight-line graphs, like the one on the previous page, are called **linear relationships**. Notice that as the number of songs increases by 2, the time in minutes increases by 1.

	+2	+2	+2	+2	
Number of Songs, y	0	2	4	6	8
Time (minutes), x	0	1	2	3	4
	+1	+1	+1	+1	

Rate of Change

$\frac{2}{1}$ = 2 songs per minute

The rate of change between any two points in a linear relationship is the same or *constant*. A linear relationship has a **constant rate of change**.

Example

Tutor

1. The balance in an account after several transactions is shown. Is the relationship between the balance and number of transactions linear? If so, find the constant rate of change. If not, explain your reasoning.

Number of Transactions	Balance ($)
3	170
6	140
9	110
12	80

+3 (, −30
+3 (, −30
+3 (, −30

As the number of transactions increases by 3, the balance in the account decreases by $30.

Since the rate of change is constant, this is a linear relationship. The constant rate of change is $\frac{-30}{3}$ or −$10 per transaction. This means that each transaction involved a $10 *withdrawal*.

Got it? Do these problems to find out.

a. _____

b. _____

Show your work.

a.

Cooling Water	
Time (min)	**Temperature (°F)**
5	95
10	90
15	86
20	82

b.

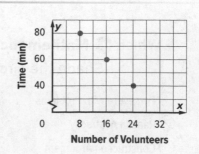

Number of Volunteers

Proportional Linear Relationships

Words   Two quantities *a* and *b* have a proportional linear relationship if they have a constant ratio and a constant rate of change.

Symbols $\frac{b}{a}$ is constant and $\frac{\text{change in } b}{\text{change in } a}$ is constant.

To determine if two quantities are proportional, compare the ratio $\frac{b}{a}$ for several pairs of points to determine if there is a constant ratio.

Example

2. **Use the table to determine if there is a proportional linear relationship between a temperature in degrees Fahrenheit and a temperature in degrees Celsius. Explain your reasoning.**

+5 +5 +5 +5

Degrees Celsius	0	5	10	15	20
Degrees Fahrenheit	32	41	50	59	68

+9 +9 +9 +9

Constant Rate of Change

$$\frac{\text{change in °F}}{\text{change in °C}} = \frac{9}{5}$$

> **Proportional Relationships**
>
> Two quantities are proportional if they have a constant ratio.

Since the rate of change is constant, this is a linear relationship.

To determine if the two scales are proportional, express the relationship between the degrees for several columns as a ratio.

$\dfrac{\text{degrees Fahrenheit}}{\text{degrees Celsius}} \longrightarrow \dfrac{41}{5} = 8.2 \qquad \dfrac{50}{10} = 5 \qquad \dfrac{59}{15} \approx 3.9$

Since the ratios are not the same, the relationship between degrees Fahrenheit degrees Celsius is *not* proportional.

Check: Graph the points on the coordinate plane. Then connect them with a line.

The points appear to fall in a straight line so the relationship is linear. ✓

The line connecting the points does not pass through the origin so the relationship is not proportional. ✓

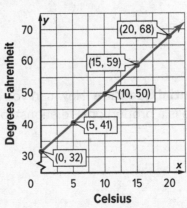

c. _____

Got it? Do this problem to find out.

c. Use the table to determine if there is a proportional linear relationship between mass of an object in kilograms and the weight of the object in pounds. Explain your reasoning.

Weight (lb)	20	40	60	80
Mass (kg)	9	18	27	36

Guided Practice

1. The amount of paint *y* needed to paint a certain amount of chairs *x* is shown in the table. Is the relationship between the two quantities linear? If so, find the constant rate of change. If not, explain your reasoning. (Example 1)

_____

_____

_____

Paint Needed for Chairs

Chairs, *x*	Cans of Paint, *y*
5	6
10	12
15	18

2. The altitude *y* of a certain airplane after a certain number of minutes *x* is shown in the graph. Is the relationship linear? If so, find the constant rate of change. If not, explain your reasoning. (Example 1)

_____

_____

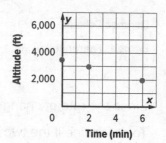

3. Determine whether a proportional relationship exists between the two quantities shown in Exercise 1. Explain your reasoning. (Example 2)

_____

_____

_____

4. **Building on the Essential Question** How can you use a table to determine if there is a proportional relationship between two quantities? _____

_____

_____

Rate Yourself!

Are you ready to move on?
Shade the section that applies.

For more help, go online to access a Personal Tutor.

Independent Practice

Go online for Step-by-Step Solutions

Determine whether the relationship between the two quantities shown in each table or graph is linear. If so, find the constant rate of change. If not, explain your reasoning. (Example 1)

1

Cost of Electricity to Run Personal Computer	
Time (h)	Cost (¢)
5	15
8	24
12	36
24	72

Show your work.

2.

Distance Traveled by Falling Object				
Time (s)	1	2	3	4
Distance (m)	4.9	19.6	44.1	78.4

3.

Italian Dressing Recipe				
Oil (c)	2	4	6	8
Vinegar (c)	$\frac{3}{4}$	$1\frac{1}{2}$	$2\frac{1}{4}$	3

4.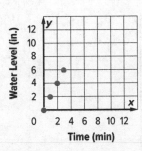

5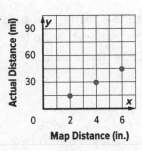

6.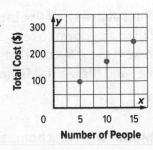

Determine whether a proportional relationship exists between the two quantities shown in the following Exercises. Explain your reasoning. (Example 2)

7. Exercise 1

8. Exercise 3

9. Exercise 5

10. **Use Math Tools** Match each table with its rate of change.

2.4 ft/min

Time (minutes)	20	30	40
Altitude (feet)	170	162	154

10 ft/min

Time (minutes)	1	2	3
Distance (feet)	20	30	40

−0.8 ft/min

Time (minutes)	4	6	8
Height (feet)	1	1.5	2

0.25 ft/min

Time (minutes)	5	10	15
Depth (feet)	12	24	36

H.O.T. Problems Higher Order Thinking

11. **Persevere with Problems** A dog starts walking, slows down, and then sits down to rest. Sketch a graph of the situation to represent the different rates of change. Label the *x*-axis "Time" and the *y*-axis "Distance".

12. **Model with Mathematics** Describe a situation with two quantities that have a proportional linear relationship.

_____

_____

_____

13. **Justify Conclusions** Each table shows a relationship with a constant rate of change. Is each relationship proportional? Justify your reasoning.

a.

Cost of Play Tickets ($)			
t	1	2	3	4
c	3.50	4.00	4.50	5.00

b.

Cost of Play Tickets ($)			
t	1	2	3	4
c	2.50	5.00	7.50	10.00

_____

_____

_____

Extra Practice

Determine whether the relationship between the two quantities shown in each table is linear. If so, find the constant rate of change. If not, explain your reasoning.

14.

Homework Help →

Sale Price Comparison	
Retail Price ($)	**Sale Price ($)**
0	0
10	5
20	10
30	15
40	20
50	25
60	30

+10 for Retail Price, +5 for Sale Price between each row.

Yes; the rate of change between the sale price and retail price is a constant value of $\frac{1}{2}$.

15.

Total Number of Customers Helped at Jewelry Store	
Time (h)	**Total Helped**
1	12
2	24
3	36
4	60

_____

_____

_____

16. Determine whether a proportional relationship exists between the two quantities in Exercise 14. Explain your reasoning. _____

_____

_____

MP Reason Abstractly Find the constant rate of change for each graph and interpret its meaning.

17.

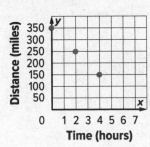

18.

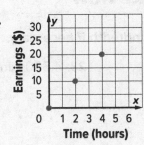

19.

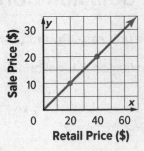

_____

_____

_____

_____

_____

_____

20. The table shows the amount of money in Will's savings account. Graph the points on the coordinate plane and connect them with a straight line.

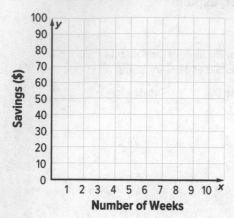

Week	Savings ($)
1	40
2	55
3	70
4	85
5	100

What is the constant rate of change? ☐

21. The graph shows the distance Bianca traveled on her 2-hour bike ride. Determine if each statement is true or false.

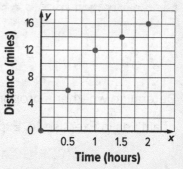

a. She traveled at a constant speed of 12 miles per hour for the entire ride.
☐ True   ☐ False

b. She traveled at a constant speed of 12 miles per hour for the first hour.
☐ True   ☐ False

c. She traveled at a constant speed of 4 miles per hour for the last hour.
☐ True   ☐ False

Common Core Spiral Review

Find the unit rate. Round to the nearest hundredth if necessary. 7.RP.1

22. 60 miles on 2.5 gallons _____

23. 4,500 kilobytes in 6 minutes _____

24. 10 red peppers for $5.50 _____

25. 72.6 meters in 11 seconds _____

Inquiry Lab
Graphing Technology: Rate of Change

 Inquiry **HOW can you use a graphing calculator to determine the rate of change?**

 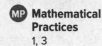 **Content Standards**
8.EE.5

MP Mathematical Practices
1, 3

At the school store, tickets to the football game are sold for $5 each. The equation $y = 5x$ can be used to find the total cost y of any number of tickets x. Find the rate of change.

What do you know? _____

What do you need to find? _____

Hands-On Activity

Recall that a rate of change is a rate that describes how one quantity changes in relation to another.

Step 1 Enter the equation. Press [Y=] 5 [X,T,θ,n].

Step 2 Graph the equation in the standard viewing window. Press [Zoom] 6.

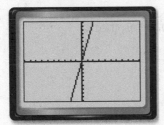

Step 3 Press [2nd] [TblSet] [▼] [▼] [ENTER] [▼] [ENTER] to generate the table automatically. Press [2nd] [Table] to access the table. Choose any two points on the line and find the rate of change.

$$\frac{\text{change in total cost}}{\text{change in number of tickets}} = \frac{\$\left(\boxed{} - \boxed{}\right)}{\left(\boxed{} - \boxed{}\right) \text{ tickets}}$$

$$= \frac{\boxed{}}{\boxed{} \text{ ticket}}$$

So, the rate of change, or unit rate, is _____.

Investigate

Work with a partner. School T-shirts are sold for $10 each and packages of markers are sold for $2.50 each.

1. For each item, write an equation that can be used to find the

   total cost y of x items. _____

2. Graph the equations in the same window as the equation from the Activity. Copy your calculator screen on the blank screen shown.

Analyze and Reflect

3. Refer to Exercises 1 and 2. Find each rate of change. Is there a relationship between the steepness of the lines on the graph and the rates of change?

   Explain. _____

   _____

   _____

   _____

On Your Own

Create

4. **MP Reason Inductively** Without graphing, write the equation of a line that is steeper than $y = \frac{1}{3}x$. Explain your reasoning.

   _____

   _____

   _____

5. **Inquiry** HOW can you use a graphing calculator to determine the rate of change?

   _____

   _____

   _____

Slope

Vocabulary Start-Up

The term *slope* is used to describe the steepness of a straight line. **Slope** is the ratio of the **rise**, or vertical change, to the **run** or horizontal change.

Complete the graphic organizer.

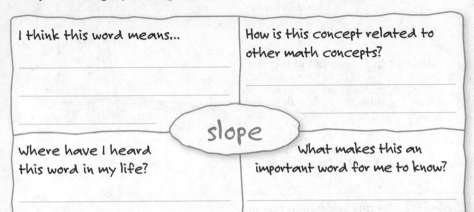

| I think this word means... | How is this concept related to other math concepts? |
| Where have I heard this word in my life? | What makes this an important word for me to know? |

slope

Essential Question

WHY are graphs helpful?

Vocabulary

slope
rise
run

Common Core State Standards

Content Standards
Preparation for 8.EE.5

MP Mathematical Practices
1, 3, 4

Real-World Link

A ride at an amusement park rises 8 feet every horizontal change of 2 feet. How could you determine the slope of the ride?

Which **MP** Mathematical Practices did you use?
Shade the circle(s) that applies.

① Persevere with Problems

⑤ Use Math Tools

② Reason Abstractly

⑥ Attend to Precision

③ Construct an Argument

⑦ Make Use of Structure

④ Model with Mathematics

⑧ Use Repeated Reasoning

Find Slope Using a Graph or Table

Slope is a rate of change. It can be positive (slanting upward) or negative (slanting downward).

slope = $\dfrac{\text{rise}}{\text{run}}$ ◄── vertical change between any two points
◄── horizontal change between the same two points

 ## Example

1. **Find the slope of the treadmill.**

slope = $\dfrac{\text{rise}}{\text{run}}$ Definition of slope

   = $\dfrac{\textbf{10 in.}}{\textbf{48 in.}}$ rise = 10 in., run = 48 in.

   = $\dfrac{5}{24}$ Simplify.

The slope of the treadmill is $\dfrac{5}{24}$.

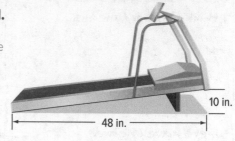

10 in.

48 in.

Show your work.

Got it? Do this problem to find out.

a. A hiking trail rises 6 feet for every horizontal change of 100 feet. What is the slope of the hiking trail?

 a. _____

Examples

Translating Rise and Run

up	→ positive
down	→ negative
right	→ positive
left	→ negative

2. **The graph shows the cost of muffins at a bake sale. Find the slope of the line.**

Choose two points on the line. The vertical change is 2 units and the horizontal change is 1 unit.

slope = $\dfrac{\text{rise}}{\text{run}}$ Definition of slope

   = $\dfrac{2}{1}$ rise = 2, run = 1

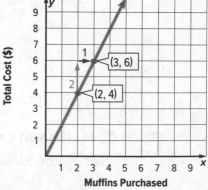

The slope of the line is $\dfrac{2}{1}$ or 2.

3. The table shows the number of pages Garrett has left to read after a certain number of minutes. The points lie on a line. Find the slope of the line.

Time (min), x	Pages left, y
1	12
3	9
5	6
7	3

Choose any two points from the table to find the changes in the x- and y-values.

$$\text{slope} = \frac{\text{change in } y}{\text{change in } x}$$ Definition of slope

$$= \frac{9 - 12}{3 - 1}$$ Use the points (1, 12) and (3, 9).

$$= \frac{-3}{2} \text{ or } -\frac{3}{2}$$ Simplify.

To check, choose two different points from the table and find the slope.

Check $$\text{slope} = \frac{\text{change in } y}{\text{change in } x}$$

$$= \frac{3 - 6}{7 - 5}$$

$$= \frac{-3}{2} \text{ or } -\frac{3}{2} \ ✓$$

> **Slope**
> In linear relationships, no matter which two points you choose, the slope, or rate of change, of the line is always constant.

Got it? Do these problems to find out.

Find the slope of each line.

b.

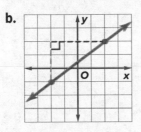

c.

x	−6	−2	2	6
y	−2	−1	0	1

Show your work.

b. _____

c. _____

Slope Formula

Key Concept

Words The slope *m* of a line passing through points (x_1, y_1) and (x_2, y_2) is the ratio of the difference in the *y*-coordinates to the corresponding difference in the *x*-coordinates.

Model

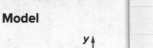
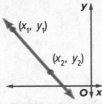

Symbols $m = \dfrac{y_2 - y_1}{x_2 - x_1}$, where $x_2 \neq x_1$

It does not matter which point you define as (x_1, y_1) and (x_2, y_2). However the coordinates of both points must be used in the same order.

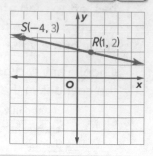

Example

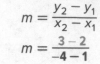

4. Find the slope of the line that passes through $R(1, 2)$, $S(-4, 3)$.

$$m = \frac{y_2 - y_1}{x_2 - x_1}$$ Slope formula

$$m = \frac{3 - 2}{-4 - 1}$$ $(x_1, y_1) = (1, 2)$
$(x_2, y_2) = (-4, 3)$

$$m = \frac{1}{-5} \text{ or } -\frac{1}{5}$$ Simplify.

Using the Slope Formula

To check Example 4, let $(x_1, y_1) = (-4, 3)$ and $(x_2, y_2) = (1, 2)$. Then find the slope.

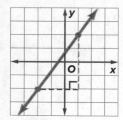
Show your work.

d. _____

e. _____

Got it? Do these problems to find out.

d. $A(2, 2)$, $B(5, 3)$ **e.** $J(-7, -4)$, $K(-3, -2)$

Guided Practice

1. Find the slope of the storage shed's roof. (Example 1)

_____

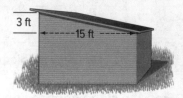
3 ft
15 ft

Find the slope of each line. (Examples 2 and 3)

2. _____

3. _____

x	0	1	2	3
y	1	3	5	7

Find the slope of the line that passes through each pair of points. (Example 4)

4. $A(-3, -2)$, $B(5, 4)$ _____   **5.** $E(-6, 5)$, $F(3, -3)$ _____

6. **Building on the Essential Question** In any linear relationship, explain why the slope is always the same.

_____

_____

_____

Rate Yourself!

How well do you understand slope? Circle the image that applies.

Clear   Somewhat   Not So
          Clear   Clear

For more help, go online to access a Personal Tutor.

Tutor

Independent Practice

Go online for Step-by-Step Solutions   eHelp

1 Find the slope of a ski run that descends 15 feet for every horizontal change of 24 feet. (Example 1)

_____

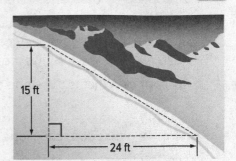
15 ft
24 ft

Find the slope of each line. (Example 2)

2.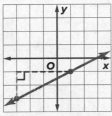

_____

3.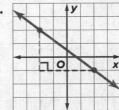

_____

The points given in the table lie on a line. Find the slope of each line.
(Example 3)

4.

x	0	2	4	6
y	9	4	−1	−6

_____

5.

x	0	1	2	3
y	3	5	7	9

_____

Find the slope of the line that passes through each pair of points. (Example 4)

6. A(0, 1), B(2, 7) _____

7 C(2, 5), D(3, 1) _____

8. E(1, 2), F(4, 7) _____

9. **MP** **Justify Conclusions** Wheelchair ramps for access to public buildings are allowed a maximum of one inch of vertical increase for every one foot of horizontal distance. Would a ramp that is 10 feet long and 8 inches tall meet this guideline? Explain your reasoning to a classmate.

_____

_____

10. **MP** **Multiple Representations** For working 3 hours, Sofia earns $30.60. For working 5 hours, she earns $51. For working 6 hours, she earns $61.20.

a. **Graphs** Graph the information with hours on the horizontal axis and money earned on the vertical axis. Draw a line through the points.

b. **Numbers** What is the slope of the line?

_____

c. **Words** What does the slope of the line represent?

How does the slope relate to the unit rate? _____

_____

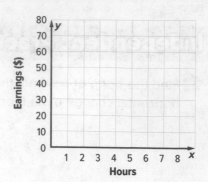

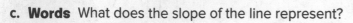

H.O.T. Problems Higher Order Thinking

11. **MP** **Find the Error** Jacob is finding the slope of the line that passes through $X(0, 2)$ and $Y(4, 3)$. Circle his mistake and correct it.

_____

_____

_____

_____

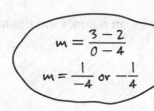

$$m = \frac{3 - 2}{0 - 4}$$

$$m = \frac{1}{-4} \text{ or } -\frac{1}{4}$$

12. **MP** **Persevere with Problems** Two lines that are parallel have the same slope. Determine whether quadrilateral $ABCD$ is a parallelogram. Justify your reasoning.

_____

_____

_____

_____

_____

_____

13. **MP** **Model with Mathematics** Give three points that lie on a line with each of the following slopes.

a. 5 _____

b. $\frac{1}{5}$ _____

c. −5 _____

Extra Practice

14. Find the slope of a road that rises 12 feet for every horizontal change of 100 feet.

$\frac{3}{25}$ _____

15. Wyatt is flying a kite in the park. The kite is a horizontal distance of 24 feet from Wyatt's position and a vertical distance of 72 feet.

Find the slope of the kite string. _____

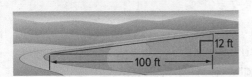

$slope = \frac{rise}{run}$ Definition of slope

Homework Help ➤ $= \frac{12\ ft}{100\ ft}$ rise = 12 ft, run = 100 ft

$= \frac{3}{25}$ Simplify.

Find the slope of each line.

16.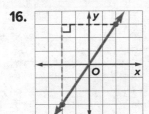

17.

MP Use Math Tools The points given in the table lie on a line. Find the slope of each line.

18.

x	−3	3	9	15
y	−3	1	5	9

_____

19.

x	−2	−1	1	2
y	−4	−2	2	4

_____

Find the slope of the line that passes through each pair of points.

20. M(−2, 3), N(7, −4) _____

21. G(−6, −1), H(4, 1) _____

22. J(−9, 3), K(2, 1) _____

23. Line *AB* represents a steep hill.

The coordinates of point *A* are ☐ and the coordinates of point *B* are ☐. The slope of the hill is ☐.

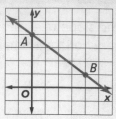

24. Lionel charted the growth of his puppy for several weeks and plotted the values on a graph. Draw a line that passes through the points (2, 4) and (10, 20).

What is the slope of the line? What does this slope represent?

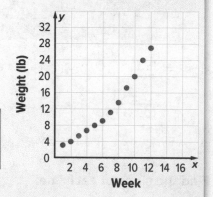

Common Core Spiral Review

25. The wait time to ride the Thunder boats is 30 minutes when 180 people are in line. Write and solve a proportion to find the wait time when 240 people are in line. **7.RP.2, 7.RP.3**

Solve each proportion. 7.RP.2, 7.RP.3

26. $\dfrac{5}{7} = \dfrac{a}{35}$

27. $\dfrac{12}{p} = \dfrac{36}{45}$

28. $\dfrac{3}{9} = \dfrac{21}{k}$

29. $\dfrac{n}{15} = \dfrac{17}{34}$

30. $\dfrac{-7}{10} = \dfrac{3.5}{j}$

31. $\dfrac{12}{18} = \dfrac{-40}{x}$

Equations in $y = mx$ Form

 Real-World Link

Charity The amount of money David can raise for the Wish Upon A Rainbow Bike-a-thon is shown in the table.

Biking Time (h), x	Money Raised ($), y
2	20
4	40
6	60

Recall that when the ratio of two variable quantities is constant, a proportional relationship exists. This relationship is called a **direct variation**. The constant ratio is called the **constant of variation** or **constant of proportionality**.

Complete the steps below to derive the equation for a direct variation.

$$\frac{\boxed{}}{\boxed{}} = \boxed{} \qquad \text{Slope formula}$$

$$\frac{y - 0}{x - 0} = m \qquad \begin{array}{l}(x_1, y_1) = (0, 0)\\(x_2, y_2) = (x, y)\end{array}$$

$$\frac{\boxed{}}{\boxed{}} = m \qquad \text{Simplify.}$$

$$y = \boxed{}\boxed{} \qquad \text{Multiplication Property of Equality}$$

1. Use the table to find the rate of change. Then write an equation in $y = mx$ form to represent the situation.

Which MP Mathematical Practices did you use?
Shade the circle(s) that applies.

① Persevere with Problems

② Reason Abstractly

③ Construct an Argument

④ Model with Mathematics

⑤ Use Math Tools

⑥ Attend to Precision

⑦ Make Use of Structure

⑧ Use Repeated Reasoning

 Essential Question

WHY are graphs helpful?

Vocab **Vocabulary**

direct variation
constant of variation
constant of proportionality

Common Core State Standards

Content Standards
8.EE.5, 8.EE.6, 8.F.2, 8.F.4

MP Mathematical Practices
1, 3, 4

Direct Variation

$y = mx$

In a direct variation equation $y = mx$, m represents the constant of variation, the constant of proportionality, the slope, and the unit rate.

Words A linear relationship is a direct variation when the ratio of y to x is a constant, m. We say y varies directly with x.

Symbols $m = \dfrac{y}{x}$ or $y = mx$, where m is the constant of variation and $m \neq 0$

Example $y = 3x$

Graph

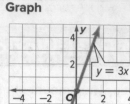

The slope of the graph of $y = mx$ is m. Since $(0, 0)$ is one solution of $y = mx$, the graph of a direct variation always passes through the origin.

Example

Tutor

1. **The amount of money Robin earns while babysitting varies directly with the time as shown in the graph. Determine the amount that Robin earns per hour.**

To determine the amount Robin earns per hour, or the unit rate, find the constant of variation.

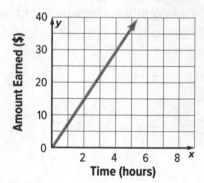

Use the points $(2, 15)$, $(3, 22.5)$, and $(4, 30)$.

$$\dfrac{\text{amount earned}}{\text{time}} \rightarrow \dfrac{15}{2} \text{ or } \dfrac{7.5}{1} \qquad \dfrac{22.5}{3} \text{ or } \dfrac{7.5}{1} \qquad \dfrac{30}{4} \text{ or } \dfrac{7.5}{1}$$

So, Robin earned $7.50 for each hour she babysits.

Got it? Do this problem to find out.

a. Two minutes after a skydiver opens his parachute, he has descended 1,900 feet. After 5 minutes, he descended 4,750 feet. If the distance varies directly with the time, at what rate is the skydiver moving?

Show your work.

a. _____

Example

Tutor

2. A cyclist can ride 3 miles in 0.25 hour. Assume that the distance biked in miles *y* varies directly with time in hours *x*. This situation can be represented by $y = 12x$. Graph the equation. How far can the cyclist ride per hour?

Make a table of values. Then graph the equation $y = 12x$. In a direct variation equation, *m* represents the slope. So, the slope of the line is $\frac{12}{1}$.

Hours, *x*	$y = 12x$	Miles, *y*
0	$y = 12(0)$	0
1	$y = 12(1)$	12
2	$y = 12(2)$	24

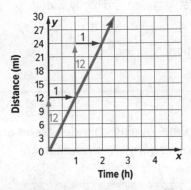

The unit rate is the slope of the line. So, the cyclist can ride 12 miles per hour.

Got it? Do this problem to find out.

b. A grocery store sells 6 oranges for $2. Assume that the cost of the oranges varies directly with the number of oranges. This situation can be represented by $y = \frac{1}{3}x$. Graph the equation. What is the cost per orange?

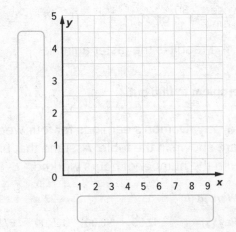

Show your work.

b. _____

Compare Direct Variations

You can use tables, graphs, words, or equations to represent and compare proportional relationships.

Table

x	15	20	25	30
y	3	4	5	6

Graph

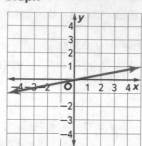

Words   y varies directly with x

Equation $y = \frac{1}{5}x$

Work Zone

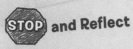

STOP and Reflect

In a proportional relationship, how is the unit rate represented on a graph? Explain below.

When the x-value changes by an amount A, the y-value will change by the corresponding amount mA.

Example

3. The distance y in miles covered by a rabbit in x hours can be represented by the equation **y = 35x**. The distance covered by a grizzly bear is shown on the graph. Which animal is faster? Explain.

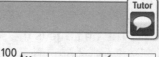

Rabbit $y = 35x$

The slope or unit rate is 35 mph.

Grizzly Bear   Find the slope of the graph.

$$\frac{rise}{run} = \frac{30}{1} \text{ or } 30$$

Since 35 > 30, the rabbit is the faster animal.

Got it? Do this problem to find out.

 Show your work.

c. _____

c. **Financial Literacy** Damon's earnings for four weeks from a part time job are shown in the table. Assume that his earnings vary directly with the number of hours worked.

Time Worked (h)	15	12	22	9
Total Pay ($)	112.50	90.00	165.00	67.50

He can take a job that will pay him $7.35 per hour worked. Which job has the better pay? Explain.

Example

4. A 3-year-old dog is often considered to be 21 in human years. Assume that the equivalent age in human years y varies directly with its age as a dog x. Write and solve a direct variation equation to find the human-year age of a dog that is 6 years old.

Let x represent the dog's actual age and let y represent the human-equivalent age.

$y = mx$	Direct variation
$21 = m(3)$	$y = 21, x = 3$
$7 = m$	Simplify.
$y = 7x$	Replace m with 7.

You want to know the human-year age or y-value when the dog is 6 years old.

$y = 7x$	Write the equation.
$y = 7 \cdot 6$	$x = 6$
$y = 42$	Simplify.

So, when a dog is 6 years old, the equivalent age in human years is 42.

Check
Graph the equation $y = 7x$.
The y-value when $x = 6$ is 42. ✔

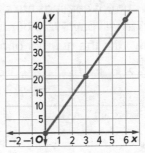

Got it? Do these problems to find out.

d. A charter bus travels 210 miles in $3\frac{1}{2}$ hours. Assume the distance traveled is directly proportional to the time traveled. Write and solve a direct variation equation to find how far the bus will travel in 6 hours.

e. A Monarch butterfly can fly 93 miles in 15 hours. Assume the distance traveled is directly proportional to the time traveled. Write and solve a direct variation equation to find how far the Monarch butterfly will travel in 24 hours.

Show your work.

d. _____

e. _____

1. A color printer can print 36 pages in 3 minutes and 108 pages in 9 minutes. If the number of pages varies directly with the time, at what rate is the color printer printing? (Example 1)

_____

2. A new compact car can travel 288 miles on nine gallons of gas. The distance driven in miles y varies directly with the number of gallons of gas x. This situation can be represented by the equation $y = 32x$. (Examples 2 and 3)

   a. Graph the equation on the coordinate plane shown.

   b. How many miles per gallon does the car get?

   _____

   c. The distance y traveled by a hybrid car using x gallons of gas can be represented by $y = 42x$. Which car gets better gas mileage? Explain.

   _____

   _____

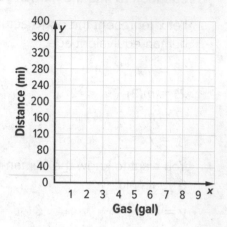

3. **Financial Literacy** Annie's current earnings are shown in the table. She was offered a new job that will pay $7.25 per hour. Assume that her earnings vary directly with the number of hours worked.

   Which job pays more an hour? (Example 3) _____

   _____

Hours, x	Money Earned ($), y
2	13.00
3	19.50
4	26.00
5	32.50

4. The height of a wide-screen television screen varies directly with its width. A television screen that is 60 centimeters wide and 33.75 centimeters high. Write and solve a direct variation equation to find the height of a television screen that is 90 centimeters wide.

   (Example 4) _____

Rate Yourself!

How well do you understand direct variation? Circle the image that applies.

Clear   Somewhat   Not So
          Clear   Clear

5. 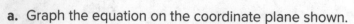 **Building on the Essential Question** What is the relationship among the unit rate, slope, and constant rate of change of a proportional linear relationship?

   _____

   _____

For more help, go online to access a Personal Tutor.

Tutor

Independent Practice

Go online for Step-by-Step Solutions

1. Dusty's earnings vary directly with the number of papers he delivers. The relationship is shown in the graph below. Determine the amount that Dusty earns for each paper he delivers. (Example 1)

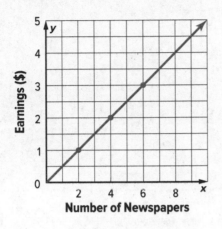

Earnings ($) / **Number of Newspapers**

2. The Thompson family is buying a car that can travel 70 miles on two gallons of gas. Assume that the distance traveled in miles y varies directly with the amount of gas used x. This can be represented by $y = 35x$. Graph the equation on the coordinate plane. How many miles does the car get per gallon

of gas? (Example 2) _____

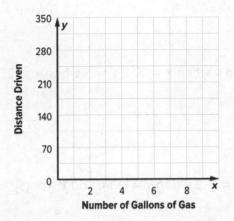

Distance Driven / **Number of Gallons of Gas**

3 Tom was comparing computer repair companies. The cost y for Computer Access for x hours is shown in the graph. The cost for Computers R Us can be represented by the equation $y = 23.5x$. Which company's repair price is lower? Explain. (Example 3)

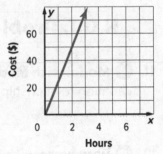

Cost ($) / **Hours**

_____

_____

4. The weight of an object on Mars varies directly with its weight on Earth. An object that weighs 50 pounds on Mars weighs 150 pounds on Earth. If an object weighs 120 pounds on Earth, write and solve a direct variation equation to find how much an object would weigh on Mars. (Example 4)

_____

Determine whether each linear function is a direct variation. If so, state the constant of variation. If not, explain why not.

5.

Pictures, x	5	6	7	8
Profit, y	20	24	28	32

_____

6.

Age, x	10	11	12	13
Grade, y	5	6	7	8

_____

7 The number of centimeters varies directly with the number of inches. Find the measure of an object in centimeters if it is 50 inches long. _____

Inches, x	6	9	12	15
Centimeters, y	15.24	22.86	30.48	38.10

MP Persevere with Problems If y varies directly with x, write an equation for the direct variation. Then find each value.

8. If $y = -12$ when $x = 9$, find y when $x = -4$. _____

9. Find y when $x = 10$ if $y = 8$ when $x = 20$. _____

10. If $y = -6$ when $x = -14$, find x when $y = -4$. _____

H.O.T. Problems Higher Order Thinking

11. **MP Model with Mathematics** Write three ordered pairs for a direct variation relationship where $y = 12$ when $x = 16$.

_____

12. **MP Persevere with Problems** The amount of stain needed to cover a wood surface is directly proportional to the area of the surface. If 3 pints are required to cover a square deck with a side of 7 feet, how many pints of stain are needed to paint a square deck with a side of 10 feet 6 inches?

_____

13. **MP Reason Inductively** Describe two real-world quantities that have a proportional linear relationship. Explain how you could change the situation to make the relationship nonproportional.

_____

_____

_____

Extra Practice

Write and graph the direct variation equation that represents each situation.

14. Hector used 3 gallons of paint to cover 1,050 square feet and 5 gallons to paint an additional 1,750 square feet. The area covered varies directly with the amount of paint used. How many square feet will one can of paint cover?

$y = 350x$; 350 square feet per gallon

Homework Help →

$y = mx$

$1,050 = m(3)$

$350 = m$

$y = 350x$

15. Nola purchased 2.5 pounds of cheese for $10.50. Her mother purchased 3 pounds of the same cheese for $12.60. The cost of cheese varies directly with the number of pounds purchased. How much does one pound of cheese cost?

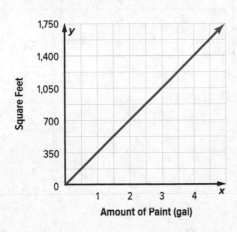

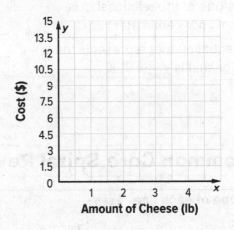

16. **STEM** When a 49 pound weight is attached to a spring, the spring stretches 7 inches. Assume that the length of the spring y varies directly with the weight attached x. Write and solve a direct variation equation to find the length of the spring when a 63 pound weight is attached.

17. **MP** **Justify Conclusions** The money raised by the Drama Club selling raffle tickets is shown in the table. They can also raise money by selling tickets to the play for $6.25 per ticket. Assume that the money raised varies directly with the number of tickets sold. Which fundraiser has the potential to raise more money? Explain your reasoning to a classmate.

Raffle Tickets Sold	25	50	75	100
Money Raised ($)	125	250	375	500

18. The table shows the amount of time a delivery truck has been driving and the distance traveled. The total distance traveled is a direct variation of the number of hours. Use the model below to find the slope.

Hours, x	2	5	7
Distance (mi), y	110	275	385

slope: $\dfrac{\boxed{} - \boxed{}}{\boxed{} - \boxed{}} = \dfrac{\boxed{}}{\boxed{}}$

Write an equation in $y = mx$ form to represent the situation. $\boxed{}$

19. Students in a science class recorded lengths of a stretched spring, as shown in the table. Determine if each statement is true or false.

Length of Stretched Spring	
Distance Stretched, x (centimeters)	Mass, y (grams)
0	0
2	12
5	30
9	54
12	72

a. The relationship represents a constant rate of change. ☐ True ☐ False

b. The slope of the relationship is 6 grams per centimeter. ☐ True ☐ False

c. The equation that represents the relationship is $y = x + 10$. ☐ True ☐ False

Common Core Spiral Review

Find the slope of each line. 8.EE.5

20. _____

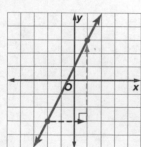

21. _____

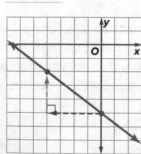

22. _____

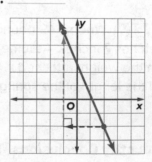

Find the slope of the line that passes through each pair of points. 8.EE.5

23. (−1, 7) and (5, 7) _____

24. (1, 3) and (1, 0) _____

25. (1, 2) and (5, 0) _____

Slope-Intercept Form

 Real-World Link

Football An interception in football is when a defensive player catches a pass made by an offensive player.

In a nonproportional linear relationship, the graph passes through the point (0, *b*) or the *y*-intercept. The **y-intercept** of a line is the *y*-coordinate of the point where the line crosses the *y*-axis.

Complete the steps to derive the equation for a nonproportional linear relationship by using the slope formula.

 = □   Slope formula

$\dfrac{y - b}{x - 0} = m$ $(x_1, y_1) = (0, b)$

$(x_2, y_2) = (x, y)$

 = m   Simplify.

$y - b = □ \cdot □$ Multiplication Property of Equality

$y = □ + □$ Addition Property of Equality

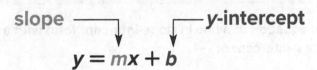

slope ⟶   ⟶ **y-intercept**

$$y = mx + b$$

How can knowing about an interception in football help you remember the definition of *y*-intercept?

_____

_____

Which MP Mathematical Practices did you use?
Shade the circle(s) that applies.

① Persevere with Problems
② Reason Abstractly
③ Construct an Argument
④ Model with Mathematics

⑤ Use Math Tools
⑥ Attend to Precision
⑦ Make Use of Structure
⑧ Use Repeated Reasoning

 Essential Question

WHY are graphs helpful?

Vocab **Vocabulary**

y-intercept
slope-intercept form

CCSS **Common Core State Standards**

Content Standards
8.EE.6, 8.F.3, 8.F.4

MP Mathematical Practices
1, 3, 4

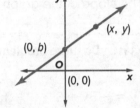

I got it!

Slope-Intercept Form of a LIne

Nonproportional linear relationships can be written in the form $y = mx + b$. This is called the **slope-intercept form**. When an equation is written in this form, m is the slope and b is the y-intercept.

Examples

Tutor

1. State the slope and the y-intercept of the graph of the equation $y = \frac{2}{3}x - 4$.

Show your work.

$y = \frac{2}{3}x + (-4)$ Write the equation in the form $y = mx + b$.

$y = mx + b$ $m = \frac{2}{3}, b = -4$

The slope of the graph is $\frac{2}{3}$, and the y-intercept is -4.

Got it? Do these problems to find out.

a. _____

b. _____

c. _____

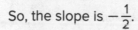

a. $y = -5x + 3$ **b.** $y = \frac{1}{4}x - 6$ **c.** $y = -x + 5$

Examples

Tutor

2. Write an equation of a line in slope-intercept form with a slope of -3 and a y-intercept of -4.

$y = mx + b$ Slope-intercept form

$y = -3x + (-4)$ Replace m with -3 and b with -4.

$y = -3x - 4$ Simplify.

3. Write an equation in slope-intercept form for the graph shown.

The y-intercept is 4. From $(0, 4)$, you move down 1 unit and right 2 units to another point on the line.

So, the slope is $-\frac{1}{2}$.

$y = mx + b$ Slope-intercept form

$y = -\frac{1}{2}x + 4$ Replace m with $-\frac{1}{2}$ and b with 4.

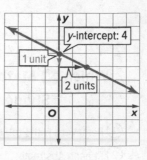

Got it? Do these problems to find out.

d. Write an equation in slope-intercept form for the graph shown.

e. Write an equation of a line in slope-intercept form with a slope of $\frac{3}{4}$ and a y-intercept of -3.

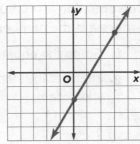

Show your work.

d. _____

e. _____

Interpret the y-Intercept

When an equation in slope-intercept form applies to a real-world situation, the slope represents the rate of change and the y-intercept represents the initial value.

 Examples

4. Student Council is selling T-shirts during spirit week. It costs $20 for the design and $5 to print each shirt. The cost y to print x shirts is given by $y = 5x + 20$. Graph $y = 5x + 20$ using the slope and y-intercept.

Step 1 Find the slope and y-intercept.

$y = 5x + 20$ slope $= 5$
             y-intercept $= 20$

Step 2 Graph the y-intercept (0, 20).

Step 3 Write the slope 5 as $\frac{5}{1}$. Use it to locate a second point on the line. Go up 5 units and right 1 unit. Then draw a line through the points.

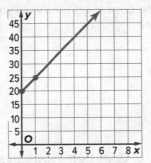

5. Interpret the slope and the y-intercept.

The slope 5 represents the cost in dollars per T-shirt. The y-intercept 20 is the one-time charge in dollars for the design.

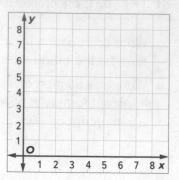

Got it? Do these problems to find out.

A taxi fare y can be determined by the equation $y = 0.50x + 3.50$, where x is the number of miles traveled.

    f. Graph the equation.

    g. Interpret the slope and the y-intercept.

Show your work

9. _____

Guided Practice

Check ✓

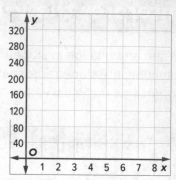

1. Liam is reading a 254-page book for school. He can read 40 pages in one hour. The equation for the number of pages he has left to read is $y = 254 - 40x$, where x is the number of hours he reads. (Examples 1, 4, and 5)

    a. State the slope and the y-intercept of the graph of the equation. _____

Show your work
    b. Graph the equation.

    c. Interpret what the slope and the y-intercept represent.

_____

_____

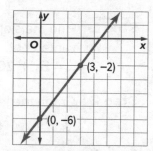

2. Write an equation in slope intercept form for the graph shown.

   (Examples 2 and 3) _____

(3, –2)

(0, –6)

3. **ⓔ Building on the Essential Question** How does the y-intercept appear in these three representations: table, equation, and graph? _____

_____

_____

_____

Rate Yourself!

How confident are you about equations in slope-intercept form? Check the box that applies.

☹ 😐 😊

☐ ☐ ☐ ☐ ☐

For more help, go online to access a Personal Tutor.

Tutor

Independent Practice

Go online for Step-by-Step Solutions

State the slope and the *y*-intercept for the graph of each equation.
(Example 1)

1. $y = 3x + 4$ _____

2. $y = -\frac{3}{7}x - \frac{1}{7}$ _____

3. $3x + y = -4$ _____

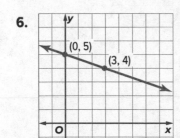

Write an equation of a line in slope-intercept form with the given slope and *y*-intercept. (Example 2)

4. slope: $-\frac{3}{4}$, *y*-intercept: -2

5. slope: $\frac{5}{6}$, *y*-intercept: 8

_____

_____

Write an equation in slope-intercept form for each graph shown. (Example 3)

6. _____

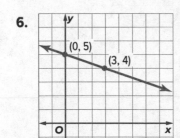

7. _____

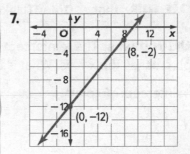

8. The Viera family is traveling from Philadelphia to Orlando for vacation. The equation $y = 1,000 - 65x$ represents the distance in miles remaining in their trip after *x* hours.
(Examples 4 and 5)

a. Graph the equation.

b. Interpret the slope and the *y*-intercept. _____

_____

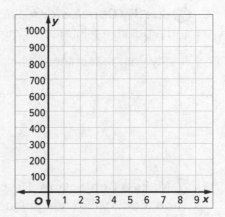

Copy and Solve. Graph each equation on a separate piece of grid paper.

9. $y = \frac{1}{3}x - 5$

10. $y = -x + \frac{3}{2}$

11. $y = -\frac{4}{3}x + 1$

12. **MP Model with Mathematics** Refer to the graphic novel frame below for Exercises a–b.

a. Write an equation in slope-intercept form for the total cost of any number of tickets at 7 tickets for $5. _____

b. Write an equation in slope-intercept form for the total cost of a wristband for all you can ride. _____

H.O.T. Problems Higher Order Thinking

13. **MP Persevere with Problems** The x-intercept is the x-coordinate of the point where a graph crosses the x-axis. What is the slope of a line that has a y-intercept but no x-intercept? Explain. _____

_____

14. **MP Reason Abstractly** Write an equation of a line that does not have a y-intercept. _____

15. **MP Justify Conclusions** Suppose the graph of a line has a negative slope and a positive y-intercept. Through which quadrants does the line pass? Justify your reasoning. _____

_____

_____

16. **MP Make a Conjecture** Describe what happens to the graph of $y = 3x + 4$ when the slope is changed to $\frac{1}{3}$.

_____

Extra Practice

State the slope and the *y*-intercept for the graph of each equation.

17. $y = -5x + 2$ _-5; 2_

18. $y = \frac{1}{2}x - 6$ _____

19. $y - 2x = 8$ _____

 Homework Help

In the equation, m = -5 and
b = 2 so the slope is -5 and
the y-intercept is 2.

**Write an equation of a line in slope-intercept form with the given slope and
y-intercept.**

20. slope: $\frac{1}{2}$; *y*-intercept: 6

21. slope: -2; *y*-intercept: 3

22. slope: $-\frac{3}{5}$; *y*-intercept: $-\frac{1}{5}$

23. **MP Persevere with Problems** The equation $y = 15x + 37$ can be
used to approximate the temperature *y* in degrees Fahrenheit based
on the number of chirps *x* a cricket makes in 15 seconds. Graph the
equation to estimate the number of chirps a cricket will make in
15 seconds if the temperature is 80°F.

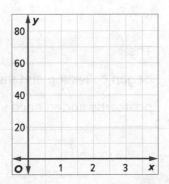

24. The Lakeside Marina charges a $35 rental fee for a boat in addition
to charging $15 an hour for usage. The total cost *y* of renting a boat
for *x* hours can be represented by the equation $y = 15x + 35$.

a. Graph the equation.

b. Interpret the slope and the *y*-intercept.

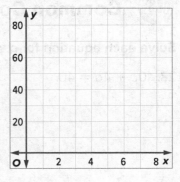

25. Write an equation in slope-intercept form for the table shown.

Number of Pizzas	0	1	2	3	4
Cost ($)	5	13	21	29	37

26. The table shows Mr. Blackwell's total earnings as a car salesman for different sale amounts.

Sales (thousands), x	$10	$20	$30
Total Earnings, y	$1,750	$3,000	$4,250

Graph the points on the coordinate plane and connect them with a straight line.

Write an equation in slope-intercept form to represent the relationship.

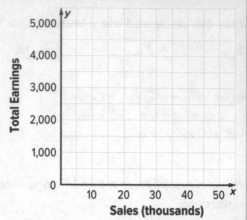

27. Jaquie has 20 postcards in her collection. Each time she goes on vacation she buys 8 postcards to add to the collection. The total number of postcards y can be represented by the equation $y = 8x + 20$. Complete the following statements regarding the line.

The slope of the line is _____ and the y-intercept is

_____ .

The _____ represents the number of postcards when she began collecting and the _____ represents the number of postcards added each vacation.

Solve each equation for d when c = 0. **7.EE.4**

28. $10c + 4d = 40$

29. $-5d = 2c + 10$

30. $-4c - 6d = 24$

Determine whether each linear relationship is proportional. If so, state the constant of proportionality. **7.RP.2**

31. _____

Pictures, x	5	6	7	8
Profit, y	20	24	28	32

32. _____

Price, x	10	15	20	25
Tax, y	0.70	1.05	1.40	1.75

Inquiry Lab
Slope Triangles

 Inquiry HOW does graphing slope triangles on the coordinate plane help you analyze them?

 Content Standards
8.EE.6

 Mathematical Practices
1, 3, 5

Donte ordered the plans shown to build a skateboard ramp. Each unit represents one foot. He wants to keep the same slope of the ramp and extend the base of the triangle three feet. How tall will the ramp be?

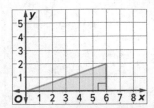

Hands-On Activity

Refer to the graph shown above. Triangle ABC is formed by the rise, run, and section of the line $y = \frac{1}{3}x$ between points A and B.

Step 1 Graph $y = \frac{1}{3}x$ on the grid paper. Draw a right triangle using the points $A(0, 0)$ and $B(6, 2)$. Label the third point C.

What is the slope of $\overline{AB}$? ☐/☐

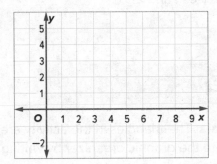

Step 2 Select any two different points on the line. Label them D and E. Draw another triangle from these two points.

Is the slope of $\overline{DE}$ the same as the slope of $\overline{AB}$? Explain.

_____

_____

Step 3 Donte wants to expand the base of the ramp 3 feet. Graph and give the coordinates of the point that will represent the extended base of the ramp. _____

Create a right triangle using the line and that point. What will be the height of the new ramp? _____

 Collaborate

Investigate

Work with a partner. Draw two right triangles for each exercise using the rise, run, and portions of the line.

1. $y = -x + 2$

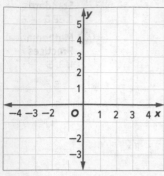

Show your work.

2. $y = x + 1$

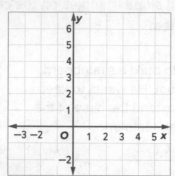

Analyze and Reflect

 Collaborate

3. (MP) **Make a Conjecture** What do you notice about the shape and size of the pair

of triangles in Exercises 1 and 2? _____

_____

Create

 On Your Own

4. (MP) **Use Math Tools** The triangles in the activity are called *slope triangles*. Complete the graphic organizer by writing three observations about slope triangles.

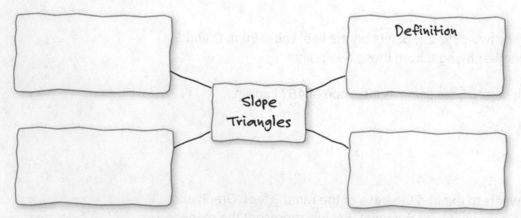

Definition

Slope Triangles

5. (inquiry) HOW does graphing slope triangles on the coordinate plane help you analyze them?

_____

_____

Lesson 5

Graph a Line Using Intercepts

Movies Mrs. Hodges spent $80 on movie tickets and drinks for her son and his friends. The total cost of x movie tickets and y drinks is represented by the equation $8x + 4y = 80$.

Item	Cost
ticket	$8
drink	$4

 Essential Question

WHY are graphs helpful?

Vocab **Vocabulary**

x-intercept
standard form

CCSS **Common Core State Standards**

Content Standards
Preparation for 8.EE.8c

MP **Mathematical Practices**
1, 3, 4

1. Complete the steps below to write the equation in slope-intercept form.

$$8x + 4y = 80$$

$$\boxed{} = \boxed{}$$

$$\frac{4y}{\boxed{}} = \frac{80 - 8x}{\boxed{}}$$

$$y = 20 - 2x$$

$$y = \boxed{}x + \boxed{}$$

slope ↑   ↑ **y-intercept**

2. Graph the equation.

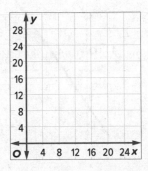

3. What does the point (0, 20) represent?

_____

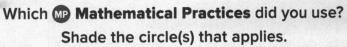

Which MP **Mathematical Practices** did you use?
Shade the circle(s) that applies.

① Persevere with Problems

② Reason Abstractly

③ Construct an Argument

④ Model with Mathematics

⑤ Use Math Tools

⑥ Attend to Precision

⑦ Make Use of Structure

⑧ Use Repeated Reasoning

Slope-Intercept Form

The **x-intercept** of a line is the x-coordinate of the point where the graph crosses the x-axis. Since any linear equation can be graphed using two points, you can use the x- and y-intercepts to graph an equation.

Example

1. State the x- and y-intercepts of $y = 1.5x - 9$. Then use the intercepts to graph the equation.

Step 1 First find the y-intercept.

$y = 1.5x + (-9)$ Write the equation in the form $y = mx + b$.

$b = -9$

Step 2 To find the x-intercept, let $y = 0$.

$0 = 1.5x - 9$ Write the equation. Let $y = 0$.

$9 = 1.5x$ Addtion Property of Equality

$\dfrac{9}{1.5} = \dfrac{1.5x}{1.5}$ Division Property of Equality

$6 = x$ Simplify.

Step 3 Graph the points (6, 0) and (0, −9) on a coordinate plane. Then connect the points.

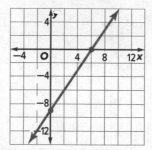

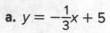 *Show your work.*

Got it? Do these problems to find out.

a. $y = -\dfrac{1}{3}x + 5$

b. $y = -\dfrac{3}{2}x + 3$

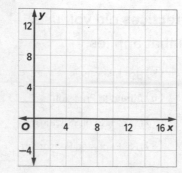

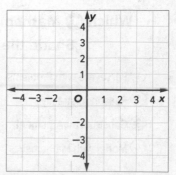

a. _____

b. _____

Standard Form

When an equation is written in the form $Ax + By = C$, where $A \geq 0$, and A, B, and C are integers, it is written in **standard form**.

 Examples

STOP and Reflect

Describe below two different methods for graphing a line.

Mauldin Middle School wants to make $4,740 from yearbooks. Print yearbooks x cost $60 and digital yearbooks y cost $15. This can be represented by the equation $60x + 15y = 4,740$.

2. Use the x- and y-intercepts to graph the equation.

To find the x-intercept, let $y = 0$. To find the y-intercept, let $x = 0$.

$60x + 15y = 4,740$

$60x + 15(0) = 4,740$

$60x = 4,740$

$x = 79$

$60x + 15y = 4,740$

$60(0) + 15y = 4,740$

$15y = 4,740$

$y = 316$

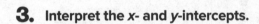

3. Interpret the x- and y-intercepts.

The x-intercept is at the point (79, 0). This means they can sell 79 print yearbooks and 0 digital yearbooks to earn $4,740.

The y-intercept is at the point (0, 316). This means they can sell 0 print yearbooks and 316 digital yearbooks to earn $4,740.

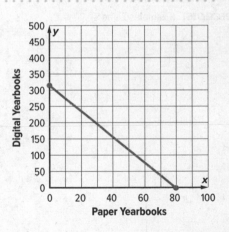

y-intercept

When an equation is written in slope-intercept form, $y = mx + b$, the y-intercept is equal to b.

Got it? Do this problem to find out.

c. Mr. Davies spent $230 on lunch for his class. Sandwiches x cost $6 and drinks y cost $2. This can be represented by the equation $6x + 2y = 230$. Use the x- and y-intercepts to graph the equation. Then interpret the intercepts.

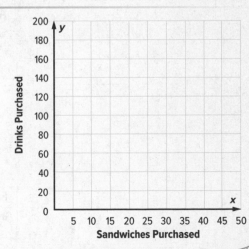

 Show your work.

c. _____

State the *x*- and *y*-intercepts of each equation. Then use the intercepts to graph the equation. (Example 1)

1. $y = 3x - 9$

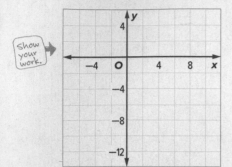

Show your work.

2. $y = \frac{1}{2}x + 2$

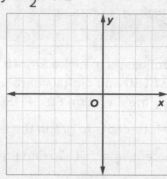

3. A store sells juice boxes in packages of 6 boxes and 8 boxes. They have 288 total juice boxes. This is represented by the function $6x + 8y = 288$. Use the *x*- and *y*-intercepts to graph the equation. Then interpret the *x*- and *y*-intercepts. (Examples 2 and 3)

_____

_____

_____

Packages with 8 Boxes

50
45
40
35
30
25
20
15
10
5
0

y

x

5 10 15 20 25 30 35 40 45 50

Packages with 6 Boxes

4. **ⓔ** **Building on the Essential Question** How can the *x*-intercept and *y*-intercept be used to graph a linear equation? _____

_____

_____

_____

Name _____ My Homework _____

Independent Practice

State the x- and y-intercepts of each equation. Then use the intercepts to graph the equation. (Example 1)

1. $y = -2x + 7$

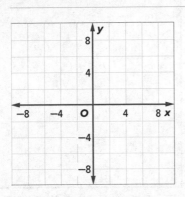

2. $y = \frac{3}{4}x + 3$

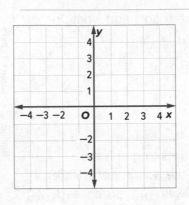

3 $12x + 9y = 15$

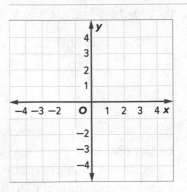

4. The table shows the cost for a clothing store to buy jeans and khakis. The total cost for Saturday's shipment, $1,800, is represented by the equation $15x + 20y = 1,800$. Use the x- and y-intercepts to graph the equation. Then interpret the x- and y-intercepts. (Examples 2 and 3)

	Jeans	Khakis
Cost per Pair ($)	15	20
Amount Shipped	x	y

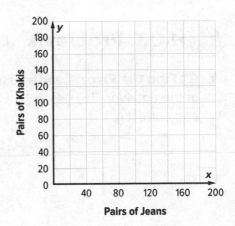

5 The total number of legs, 1,500, on four-legged and two-legged animals in a zoo can be represented by the equation $4x + 2y = 1,500$. Use the x- and y-intercepts to graph the equation. Then interpret the x- and y-intercepts. (Examples 2 and 3)

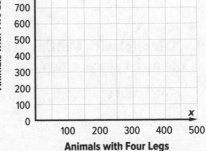

Copyright © McGraw-Hill Education

Lesson 5 Graph a Line Using Intercepts **213**

6. **MP Multiple Representations** The table shows the group rate for admission tickets for adults and children to an amusement park.

	Adult	Children
Ticket Price ($)	45	30
Tickets Purchased	x	y

a. **Symbols** The total cost of a group's tickets is $1,350. Write an equation to represent the number of adults' and children's tickets purchased.

_____

b. **Words** What are the x- and y-intercepts and what do they represent? _____

_____

_____

c. **Graphs** Use the x- and y-intercepts to graph the equation. Use the graph to find the number of children's tickets purchased if 20 adult tickets were purchased.

_____

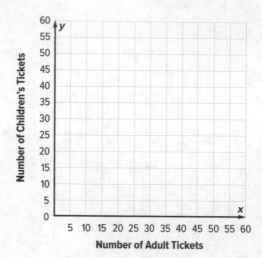

Number of Children's Tickets (y-axis)
Number of Adult Tickets (x-axis)

H.O.T. Problems Higher Order Thinking

7. **MP Find the Error** Carmen is finding the x-intercept of the equation $3x - 4y = 12$. Find her mistake and correct it.

_____

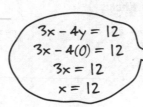

$3x - 4y = 12$
$3x - 4(0) = 12$
$3x = 12$
$x = 12$

8. **MP Persevere with Problems** The perimeter of a rectangle that is x units wide and y units long is 24 centimeters.

a. Write an equation in standard form for the perimeter. _____

b. Find the x- and y-intercepts. Does either intercept make sense as a solution for this situation? Explain. _____

_____

_____

9. **MP Model with Mathematics** Write two equations, one with an x-intercept but no y-intercept, and one with a y-intercept but no x-intercept.

x-intercept equation: _____

y-intercept equation: _____

Extra Practice

10. State the x- and y-intercepts of the equation $y = \frac{2}{3}x - \frac{1}{3}$. Then use the intercepts to graph the equation.

Find the y-intercept.

$y = \frac{2}{3}x + \left(-\frac{1}{3}\right)$

$b = -\frac{1}{3}$

Find the x-intercept.

$y = \frac{2}{3}x + \left(-\frac{1}{3}\right)$

$0 = \frac{2}{3}x + \left(-\frac{1}{3}\right)$

$\frac{1}{3} = \frac{2}{3}x$

$\left(\frac{3}{2}\right)\frac{1}{3} = \left(\frac{3}{2}\right)\frac{2}{3}x$

$\frac{1}{2} = x$

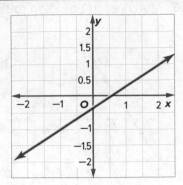

Homework Help

Copy and Solve **State the x- and y-intercepts of each equation. Then use the intercepts to graph each equation on a separate sheet of grid paper.**

11. $2x + 3y = 24$

12. $y = -\frac{8}{9}x - 16$

13. $5x + 3y = 30$

14. Tiffany has 15 teaspoons of chocolate chips. She uses $1\frac{1}{2}$ teaspoons for each muffin. The total number of teaspoons of chocolate chips that she has left y after making x muffins can be given by $y = -\frac{3}{2}x + 15$. Graph the equation. Then interpret the x- and y-intercepts. _____

_____

_____

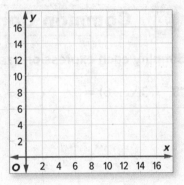

15. **MP** **Use Math Tools** Miriam has $440 to pay a painter to paint her basement. The painter charges $55 per hour. The equation $y = 440 - 55x$ represents the amount of money y she has after x number of hours worked by the painter. Graph the equation. Then interpret the x- and y-intercepts. _____

_____

_____

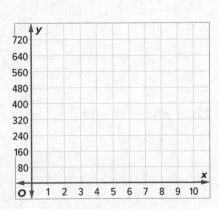

16. Match each equation to the appropriate graph below.

| $2x - 3y = -6$ |
| $3x - 2y = -6$ |
| $3x + 2y = 6$ |
| $2x + 3y = 6$ |

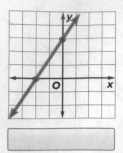 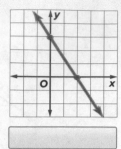 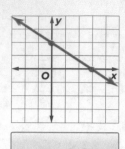 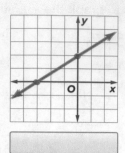

17. The equation $12x - 10y = 600$ represents the total amount Student Council spent on supplies for a school fundraiser. Fill in the boxes to make a true statement.

The *x*-intercept of the line is [ ] and the *y*-intercept is [ ].

CCSS ## Common Core Spiral Review

Simplify each expression. 7.EE.2

18. $-3(x + 6) =$ _____

19. $\frac{2}{3}(3x + 6) - 3 =$ _____

20. $4t + 10 - 5 - 3t =$ _____

21. $5x + 6 - x =$ _____

22. $-\frac{1}{4}(4x - 8) + 18 =$ _____

23. $2a + 4 - 8a - 10 =$ _____

MP Problem-Solving Investigation
Guess, Check, and Revise

CCSS **Content Standards**
8.EE.8

MP **Mathematical Practices**
1, 3, 4

Case #1 Polar Plunge

Adrienne's class is going to the zoo to see a polar bear exhibit. Student admission is $2 and adult admission is $4. They spent $66 on 30 tickets.

How many students and adults are going to the zoo?

Understand *What are the facts?*

The student cost is $2 and the adult cost is $4. There are 30 people on the trip.

Plan *What is your strategy to solve this problem?*

Make a guess and check to see if your guess is correct.

Solve *How can you apply the strategy?*

Make a table.

Students   Adults

s	a	2s + 4a	Check
26	4	2(26) + 4(4) = 68	too high
29	1	2(29) + 4(1) = 62	too low
28	2	2(28) + 4(2) = ☐	
27	3	2(27) + 4(3) = ☐	

So, 27 students and 3 adults are going to the zoo.

Check *Does the answer make sense?*

27 + 3 = 30 and 2(27) + 4(3) = 66; the guess is correct. ✓

Analyze the Strategy

MP **Justify Conclusions** Twenty-three students and 5 adults would also spend $66 to get into the zoo. Explain why this cannot be the correct solution.

Case #2 Coins

Gerardo has $2.50 in quarters, dimes, and nickels.

If he has 18 coins, how many of each coin does he have?

1 Understand

Read the problem. What are you being asked to find?

I need to find _____.

Underline key words and values in the problem. What information do you know?

There are ☐ coins that have a sum of ☐.

The coins are a combination of _____.

Is there any information that you do *not* need to know?

_____.

2 Plan

Choose a problem-solving strategy.

I will use the _____ strategy.

3 Solve

Use your problem-solving strategy to solve the problem.

Q	D	N	Sum	Number of Coins	Check

So, _____.

4 Check

Use information from the problem to check your answer.

(____ × 0.25) + (____ × 0.10) + (____ × 0.05) = $2.50; the answer is correct.

Work with a small group to solve the following cases. Show your work on a separate piece of paper.

Case #3 Wrap it Up

Shya works part-time at a gift-wrapping store. The store sells wrapping paper rolls and square packages of wrapping paper. There are a total of 125 rolls and packages. Each roll costs $3.50 and each package costs $2.25. The total cost of all of the rolls and packages is $347.50.

How many rolls of wrapping paper are there?

_____

Case #4 Family

Five siblings have a combined age of 195 years. The oldest is 13 years older than the youngest, Marc. The middle child, Anne, is five years younger than Josie. The other two siblings are 6 years apart.

If the second oldest child is 42, what are the ages of the siblings?

_____

Case #5 Sport Trading Cards

Baseball cards come in packages of 8 and 12. Brighton bought some of each type for a total of 72 baseball cards.

How many of each package did he buy?

_____

Use any strategy!

Case #6 Future Careers

One hundred fifteen students could sign up to hear three different speakers for career day. Seventy students heard the nurse speak, 52 heard the firefighter, and 78 heard the Webmaster. Some students heard more than one speaker. The results are shown in the table above.

How many students signed up only for Webmaster?

_____

Number of Students	Speaker
15	all three
20	nurse and firefighter
30	Webmaster and nurse
12	firefighter only

Mid-Chapter Check

Copyright © McGraw-Hill Education

Vocabulary Check

1. **MP** **Be Precise** Define *linear relationship*. Give an example of a linear relationship. (Lesson 1)

_____

_____

Skills Check and Problem Solving

Find the slope of the line that passes through each pair of points. (Lesson 2)

2. $A(2, 5)$, $B(3, 1)$

_____

3. $C(-1, 2)$, $D(-5, 2)$

_____

4. $E(5, 2)$, $F(2, -3)$

_____

5. $G(4, 3)$, $H(-2, -6)$

_____

6. Ernesto baked 3 cakes in $2\frac{1}{2}$ hours. Assume that the number of cakes baked varies directly with the number of hours. Write and solve a direct variation equation to find how many cakes can he bake in $7\frac{1}{2}$ hours. (Lesson 3)

_____

7. The total money y Aaron earned mowing x lawns is shown by the equation $y = 15x + 25$. What does the slope represent? (Lesson 4)

_____

8. **MP** **Persevere with Problems** A company logo has four concentric circles Suppose you graph the points (diameter, circumference) and connect them with a line. In terms of pi (π), what is the slope of the resulting line? (Lesson 4)

_____

Write Linear Equations

 Real-World Link

Zoo The cost for 1, 2, 3, and 4 people to go the zoo is shown in the table.

Number of People, x	1	2	3	4
Total Cost, y	$13	$22	$31	$40

1. Is the relationship linear? Explain.

_____

_____

2. What is the slope of the related graph? ☐

3. Choose an ordered pair. (☐ , ☐) Then substitute the values in the equation below.

$$y = m \quad x + b$$

☐ = ☐ · ☐ + b

4. Solve for b to find the y-intercept.

$b = $ ☐

5. Write an equation of the line in slope-intercept form.

_____

6. Graph the data from the table on the coordinate plane.

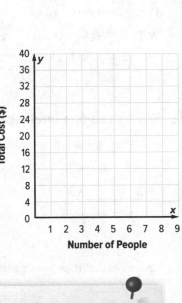

Total Cost ($)

Number of People

 Essential Question

WHY are graphs helpful?

 Vocabulary

point-slope form

Common Core State Standards

Content Standards
Preparation for 8.EE.8c

MP **Mathematical Practices**
1, 2, 3, 4, 5, 7

Which MP Mathematical Practices did you use?
Shade the circle(s) that applies.

① Persevere with Problems
② Reason Abstractly
③ Construct an Argument
④ Model with Mathematics

⑤ Use Math Tools
⑥ Attend to Precision
⑦ Make Use of Structure
⑧ Use Repeated Reasoning

Point-Slope Form of a Linear Equation

Words

The linear equation $y - y_1 = m(x - x_1)$ is written in point-slope form, where (x_1, y_1) is a given point on a nonvertical line and m is the slope of the line.

Graph

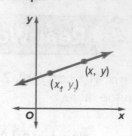

Symbols $y - y_1 = m(x - x_1)$

Slope

The point-slope form of a linear equation is tied directly to the definition of slope.

$$\frac{y - y_1}{x - x_1} = m$$

$$(y - y_1) = m(x - x_1)$$

You can write an equation of a line in slope-intercept form when you know the slope and the *y*-intercept. You can write an equation of a line in **point-slope form** when you are given the slope and the coordinates of a point on the line that is not the *y*-intercept.

Examples

Tutor

1. Write an equation in point-slope form for the line that passes through (−2, 3) with a slope of 4.

$y - y_1 = m(x - x_1)$	Point-slope form
$y - 3 = 4[x - (-2)]$	$(x_1, y_1) = (-2, 3)$, $m = 4$
$y - 3 = 4(x + 2)$	Simplify.

2. Write the slope-intercept form of the equation from Example 1.

$y - 3 = 4(x + 2)$	Write the equation.
$y - 3 = 4x + 8$	Distributive Property
$\underline{+3 \quad = \quad +3}$	Addition Property of Equality
$y = 4x + 11$	Simplify.

Check: Substitute the coordinates of the given point in the equation.

$y = 4x + 11$

$3 \overset{?}{=} 4(-2) + 11$

$3 = 3$ ✓

Show your work.

Got it? Do this problem to find out.

a. Write an equation in point-slope form and slope-intercept form for the line that passes through (−1, 2) and has a slope of $-\frac{1}{2}$.

a. _____

Write a Linear Equation

From Slope and a Point
- Substitute the slope m and the coordinates of the point in $y - y_1 = m(x - x_1)$.

From Slope and y-intercept
- Substitute the slope m and y-intercept b in $y = mx + b$.

From a Graph
- Find the y-intercept b and the slope m from the graph, then substitute the slope and y-intercept in $y = mx + b$.

From Two Points
- Use the coordinates of the points to find the slope. Substitute the slope and coordinates of one of the points in $y - y_1 = m(x - x_1)$.

From a Table
- Use the coordinates of the two points to find the slope, then substitute the slope and coordinates of one of the points in $y - y_1 = m(x - x_1)$.

The form you use to write a linear equation is based on the information you are given.

Example

Tutor

3. **Write an equation in point-slope form and slope-intercept form for the line that passes through (8, 1) and (−2, 9).**

Step 1 Find the slope.

$$m = \frac{y_2 - y_1}{x_2 - x_1} \qquad \text{Slope formula}$$

$$m = \frac{9 - 1}{-2 - 8} \qquad (x_1, y_1) = (8, 1), (x_2, y_2) = (-2, 9)$$

$$m = -\frac{8}{10} \text{ or } -\frac{4}{5} \qquad \text{Simplify.}$$

Step 2 Use the slope and the coordinates of either point to write the equation in point-slope form.

$$y - y_1 = m(x - x_1) \qquad \text{Point-slope form}$$

$$y - 1 = -\frac{4}{5}(x - 8) \qquad (x_1, y_1) = (8, 1), m = -\frac{4}{5}.$$

So, the point-slope form of the equation is $y - 1 = -\frac{4}{5}(x - 8)$.
In slope-intercept form, this is $y = -\frac{4}{5}x + \frac{37}{5}$.

Show your work.

Got it? **Do these problems to find out.**

c. (3, 0) and (6, −3)   **d.** (−1, 2) and (5, −10)

c. _____

d. _____

 Tutor

4. The cost of assistance dog training sessions is shown in the table. Write an equation in point-slope form to represent the cost y of attending x dog training sessions.

Number of Sessions	Cost ($)
5	165
10	290

Find the slope of the line. Then use the slope and one of the points to write the equation of the line.

$m = \dfrac{290 - 165}{10 - 5}$ $(x_2, y_2) = (10, 290), (x_1, y_1) = (5, 165)$

$m = \dfrac{125}{5}$ or 25   Simplify.

$y - 165 = 25(x - 5)$ Replace (x_1, y_1) with (5, 165) and m with 25 in the point-slope form equation.

So, the equation of the line is $y - 165 = 25(x - 5)$.

 Show your work.

Got it? Do this problem to find out.

e. _____

e. The cost for making spirit buttons is show in the table. Write an equation in point-slope form to represent the cost y of making x buttons.

Number of Buttons	Cost ($)
100	25
150	35

Guided Practice

 Check ✓

Write an equation in point-slope form and slope-intercept form for each line. (Examples 1–3)

1. passes through (2, 5), slope = 4

 Show your work.

2. passes through (−3, 1) and (−2, −1)

3. Janelle is planning a party. The cost for 20 people is $290. The cost for 45 people is $590. Write an equation in point-slope form to represent the cost y of having a party for x people. (Example 4)

4. 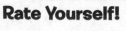 **Building on the Essential Question** How does using the point-slope form of a linear equation make it easier to write the equation of a line?

Rate Yourself!

How confident are you about writing linear equations? Check the box that applies.

For more help, go online to access a Personal Tutor.

 Tutor

Independent Practice

Write an equation in point-slope form and slope-intercept form for each line.
(Examples 1–3)

1. passes through (1, 9), slope = 2

 Show your work.
_____

2. passes through (4, −1), slope = −3

_____

3. passes through (−4, −5), slope = $\frac{3}{4}$

_____

4. passes through (3, −6) and (−1, 2)

_____

5 passes through (4, −4) and (8, −10)

_____

6. passes through (3, 4) and (5, −4)

_____

7. **STEM** For a science experiment, Mala measured the height of a plant every week. She recorded the information in the table. Assuming the growth is linear, write an equation in point-slope form to represent the height y of the plant after x weeks. (Example 4)

Weeks	Height (in.)
5	13
10	14

_____

8. After 2 seconds on a penalty kick in soccer, the ball travels 160 feet. After 2.75 seconds on the same kick, the ball travels 220 feet. Write an equation in point-slope form to represent the distance y of the ball after x seconds.

(Example 4) _____

Write each equation in standard form.

9 $y - 4 = -3(x - 3)$

10. $y + 9 = 2(x + 5)$

11. **MP Identify Structure** Draw a line connecting the form of the equation to the correct equations.

| Slope-Intercept Form |

| Standard Form |

| Point-Slope Form |

$5x + 3y = 12$

$y = 2x - 8$

$7x = y$

$y - 8 = \frac{1}{2}(x - 9)$

$4x - 6y = 24$

$y = 10 - 3x$

🔥 H.O.T. Problems Higher Order Thinking

12. **MP Reason Abstractly** Write a linear equation that is in point-slope form. Identify the slope and name a point on the line.

13. **MP Persevere with Problems** The equation of a line is $y = -\frac{1}{2}x + 6$. Write an equation in point-slope form for the same line. Explain the steps that you used.

14. **MP Persevere with Problems** Order the steps to write a linear equation in slope-intercept form if you know the slope of the line and a point on the line.

_____ Simplify the equation.

_____ Use the Distributive Property to multiply the slope by x and x_1.

_____ Substitute the slope m and the coordinates of the point (x_1, y_1) into the point-slope formula.

_____ Use the Addition Property of Equality.

Extra Practice

Write an equation in point-slope form and slope-intercept form for each line.

15. passes through $(-7, 10)$, slope $= -4$

$y - 10 = -4(x + 7); y = -4x - 18$

$y - y_1 = m(x - x_1)$
$y - 10 = -4(x + 7)$
$y - 10 = -4x - 28$
$\underline{ + 10 = + 10}$
$y = -4x - 18$

Homework Help

16. passes through $(1, 2)$ and $(3, 4)$

$y - 4 = 1(x - 3); y = x + 1$

$m = \dfrac{y_2 - y_1}{x_2 - x_1} = \dfrac{4 - 2}{3 - 1} = \dfrac{2}{2}$ or 1

$y - y_1 = m(x - x_1)$
$y - 4 = 1(x - 3)$
$y - 4 = x - 3$
$\underline{ + 4 = + 4}$
$y = x + 1$

17. passes through $(6, 2)$, slope $= \dfrac{2}{3}$

18. passes through $(2, -2)$ and $(4, -1)$

Write each equation in standard form.

19. $y + 1 = \dfrac{4}{5}(x - 3)$

20. $y - 8 = -\dfrac{1}{2}(x + 4)$

MP Use Math Tools Write the point-slope form of an equation for each line graphed.

21.

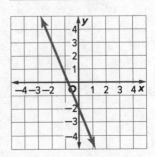

22.

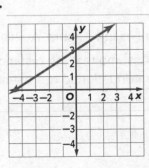

23. The table shows some ordered pairs that lie on a line. Which equations could represent the line? Select all that apply.

x	−1	0	1	2
y	−6	−2	2	6

☐ $y = 4x - 2$ ☐ $y = -4x + 1$ ☐ $y - 2 = 4(x - 1)$ ☐ $y - 2 = 4(x - 6)$

24. After 4 hours of driving, Jan is 248 miles away from home. After 6 hours of driving, she is 372 miles from home. Select the correct values to complete the model below.

1	62
4	248
6	372

slope: $\dfrac{\boxed{} - \boxed{}}{\boxed{} - \boxed{}} = \dfrac{\boxed{}}{\boxed{}}$

What is the point-slope form of the line that represents this situation?

---

25. Use the information in the table to find the constant rate of change in dollars per hour. **7.RP.2**

Time (h)	0	1	2	3
Wage ($)	0	9	18	27

26. Graph $y = 4x$. **8.EE.5**

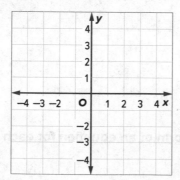

27. A train traveled 150 miles in $1\frac{1}{4}$ hours. At this rate, how far will the train travel after 5 hours? Assume that the distance traveled varies directly with the time traveled. Write an equation to represent the situation. **7.RP.2**

Inquiry Lab

Graphing Technology: Model Linear Behavior

 HOW does using technology help you to determine if situations display linear behavior?

CCSS **Content Standards** 8.F.4, 8.F.5

MP **Mathematical Practices** 1, 3, 5

Simone and Lee walked to school at about 3 miles per hour. Use the Investigation to see if the relationship between time and distance is a linear relationship.

Hands-On Activity

Step 1 Connect a motion detector to your calculator. Start the data collection program by pressing APPS (CBL/CBR), ENTER, and then select Ranger, Applications, Meters, Dist Match.

Step 2 Place the detector on a desk or table so that it can read the motion of a walker.

Step 3 Mark the floor at a distance of 1 and 6 meters from the detector. Have a partner stand at the 1-meter mark.

Step 4 When you press the button to begin collecting data, have your partner begin to walk away from the detector at a slow but steady pace.

Step 5 Stop collecting data when your partner passes the 6-meter mark.

Step 6 Press ENTER to display a graph of the data. The x-values represent equal intervals of time in seconds. The y-values represent the distances from the detector in meters.

Describe the DISTANCE graph of the data. Does the relationship between time and distance appear to be linear? Explain.

_____

_____

Inquiry Lab Graphing Technology: Model Linear Behavior **229**

Investigate

MP Use Math Tools Refer to the Activity. Work with a partner.

1. Use the ⬚TRACE feature on your calculator to find the *y*-intercept on the graph. Interpret its meaning. _____

_____

2. Press ⬚STAT 1. The time data is in ⬚L1 and the distance data is in ⬚L2 . Use these data to calculate the rate of change $\frac{distance}{time}$ for three pairs of points.

Point 1 (time, distance)	Point 2 (time, distance)	$\frac{distance_2 - distance_1}{time_2 - time_1}$	rate of change

3. **MP Justify Conclusions** Does the table in Exercise 2 support your conclusion about the graph in the Activity? Explain. _____

_____

Analyze and Reflect

4. Predict how the graph and answers to Exercise 2 would change if the person in the activity were to:

a. move at a steady but *quicker* pace *away* from the detector.

_____

b. move at a steady pace *toward* the detector.

_____

Create

5. **MP Reason Inductively** How could you change the situation to be one that does not display linear behavior? _____

6. **Inquiry** HOW does using technology help you to determine if a situation displays linear behavior?

_____

Inquiry Lab

Graphing Technology: Systems of Equations

 Inquiry **HOW can I use a graphing calculator to find one solution for a set of two equations?**

CCSS **Content Standards**
8.EE.8, 8.EE.8a, 8.EE.8b, 8.EE.8c

MP **Mathematical Practices**
1, 3, 5, 7

Web site A charges $3 plus $1 per pound to ship an item. Web site B charges $1 plus $2 per pound to ship the same item. For an object that weighs x pounds, the charges for Web site A are represented by $y = x + 3$. The charges for Web site B are represented by $y = 2x + 1$. At what point are the charges the same?

What do you know? _____

_____

What do you need to know? _____

_____

Hands-On Activity

Use a graphing calculator to generate a table of values for $y = x + 3$ and $y = 2x + 1$. Then use the table to find the total cost to ship objects that weigh 0, 1, 2, or 3 pounds.

Step 1 Press [Y=]. Then enter each equation.

Step 2 Set up the table. Press [2nd] [TblSet] to display the table setup screen. Press [▼] [▼] [▶] [ENTER] to highlight Indpnt: Ask. Then press [▼] [ENTER] to highlight Depend: Auto.

Step 3 Access the table by pressing [2nd] [Table]. Now key in your input values, pressing [ENTER] after each one. Fill in the table. The first one is done for you.

So, the point when the charges are the same

is _____ .

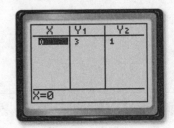

Investigate

MP Use Math Tools Work with a partner. Refer to the Activity.

1. For what number of pounds are the charges for Web site A less than those for Web site B? _____

2. For what number of pounds are the charges for Web site A greater than the ones for Web site B? _____

3. Press GRAPH. Copy your calculator screen on the blank screen shown.

4. At what point do the two lines intersect? What does this ordered pair represent? _____

_____

5. How does the point of intersection of the two lines compare to the answer to the Activity? _____

_____

Analyze and Reflect

6. **MP Use Math Tools** Use a graphing calculator to graph each set of equations in the table. Find the point of intersection of the two lines.

Set of Equations	Point of Intersection
$y = 2x + 2$ $y = x + 2$	
$y = 3x + 5$ $y = -x - 3$	

7. **MP Identify Structure** Explain what the point of intersection represents.

_____

Create

8. **MP Model with Mathematics** Describe a real-world situation that would involve finding the point of intersection of two lines.

_____

_____

_____

9. **Inquiry** HOW can I use a graphing calculator to find one solution for a set of two equations? _____

_____

Solve Systems of Equations by Graphing

Real-World Link

Activities A campground offers tubing, kayak, and bicycle rentals as shown.

	Deposit ($)	Cost per Hour ($)
Tube	15	4.20
Kayak	25	6.50
Bicycle	20	7.50

1. Write an equation to represent the total cost y of renting a tube for any number of hours x. _____

2. Write an equation to represent the total cost y of renting a kayak for any number of hours x. _____

3. Write an equation to represent the total cost y of renting a bicycle for any number of hours x. _____

4. Find the cost to rent each item for 1, 2, 3, 4, and 5 hours.

Hours	Cost of Tube ($)	Cost of Kayak ($)	Cost of Bicycle ($)
1			
2			
3			
4			
5			

Essential Question

WHY are graphs helpful?

 Vocabulary

systems of equations

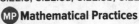 **Common Core State Standards**

Content Standards
8.EE.8, 8.EE.8a, 8.EE.8b, 8.EE.8c

MP **Mathematical Practices**
1, 3, 4, 7

Which MP **Mathematical Practices** did you use?
Shade the circle(s) that applies.

① Persevere with Problems

② Reason Abstractly

③ Construct an Argument

④ Model with Mathematics

⑤ Use Math Tools

⑥ Attend to Precision

⑦ Make Use of Structure

⑧ Use Repeated Reasoning

Systems of Equations

Two or more equations with the same set of variables are called a **system of equations**. For example, $y = 4x$ and $y = 4x + 2$ together are a system of equations.

You can estimate the solution of a system of equations by graphing the equations on the same coordinate plane. The ordered pair for the point of intersection of the graphs is the solution of the system because the point of intersection simultaneously satisfies both equations.

Example

Tutor

1. **Solve the system $y = -2x - 3$ and $y = 2x + 5$ by graphing.**

Graph each equation on the same coordinate plane.

The graphs appear to intersect at $(-2, 1)$.

Check this estimate by replacing x with -2 and y with 1.

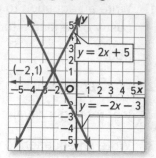

Check $\quad y = -2x - 3 \qquad y = 2x + 5$

$\qquad 1 \overset{?}{=} -2(-2) - 3 \qquad 1 \overset{?}{=} 2(-2) + 5$

$\qquad 1 = 1 \checkmark \qquad\qquad 1 = 1 \checkmark$

The solution of the system is $(-2, 1)$.

Got it? **Do these problems to find out.**

a. $y = x - 1$
$ y = 2x - 2$

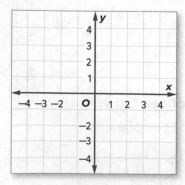

b. $y = 4x$
$ y = x + 3$

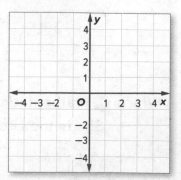

Show your work.

a. _____

b. _____

Examples

Tutor

Gregory's Motorsports has motorcycles (two wheels) and ATVs (four wheels) in stock. The store has a total of 45 vehicles, that, together, have 130 wheels.

2. **Write a system of equations that represents the situation.**

Let y represent the motorcycles and x represent the ATVs.

$y + x = 45$ The number of motorcycles and ATVs is 45.

$2y + 4x = 130$ The number of wheels equals 130.

3. **Solve the system of equations. Interpret the solution.**

Write each equation in slope-intercept form.

$x + y = 45$ $\qquad\qquad$ $2y + 4x = 130$

$\quad y = -x + 45$ $\qquad\qquad\quad$ $2y = -4x + 130$

$\qquad\qquad\qquad\qquad\qquad\quad y = -2x + 65$

Graph both equations on the same coordinate plane. The equations intersect at (20, 25).

The solution is (20, 25). This means that the store has 20 ATVs and 25 motorcycles.

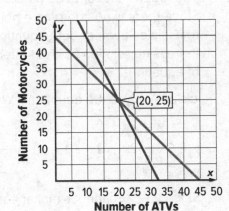

Check

$\qquad x + y = 45$ $\qquad\qquad$ $2y + 4x = 130$

$20 + 25 \overset{?}{=} 45$ $\qquad$ $2(25) + 4(20) \overset{?}{=} 130$

$\qquad 45 = 45 \checkmark$ $\qquad\qquad$ $130 = 130 \checkmark$

Got it? Do this problem to find out.

Show your work.

c. Creative Crafts gives scrapbooking lessons for $15 per hour plus a $10 supply charge. Scrapbooks Incorporated gives lessons for $20 per hour with no additional charges. Write and solve a system of equations that represents the situation. Interpret the solution.

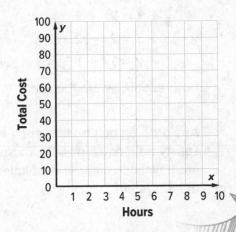

c. _____

Number of Solutions

The graph of a system of equations indicates the number of solutions.

- If the lines intersect, there is one solution.
- If the lines are parallel, there is no solution.
- If the lines are the same, there are an infinite number of solutions.

Examples

Tutor

Solve each system of equations by graphing.

4. $y = 2x + 1$
$y = 2x - 3$

Graph each equation on the same coordinate plane.

The graphs appear to be parallel lines. Since there is no coordinate point that is a solution of both equations, there is no solution for this system of equations.

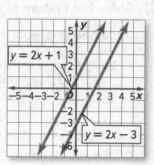

Check Analyze the equations. Write them in standard form.

$$y = 2x + 1 \qquad\qquad y = 2x - 3$$
$$y - 2x = 2x - 2x + 1 \qquad y - 2x = 2x - 2x - 3$$
$$y - 2x = 1 \qquad\qquad y - 2x = -3$$

Since $y - 2x$ cannot simultaneously be 1 and -3, there is no solution. ✓

5. $y = 2x + 1$
$y - 3 = 2x - 2$

Write $y - 3 = 2x - 2$ in slope-intercept form.

$$y - 3 = 2x - 2 \qquad \text{Write the equation.}$$
$$y - 3 + 3 = 2x - 2 + 3 \qquad \text{Add 3 to each side.}$$
$$y = 2x + 1 \qquad \text{Simplify.}$$

Both equations are the same. Graph the line.

Any ordered pair on the graph will satisfy both equations. So, there are an infinite number of solutions of the system.

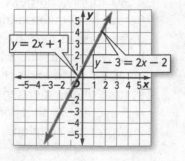

Got it? Do these problems to find out.

Solve each system of equations by graphing.

d. $y = \frac{2}{3}x + 3$

$3y = 2x + 15$

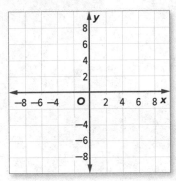

e. $y - x = 1$

$y = x - 2 + 3$

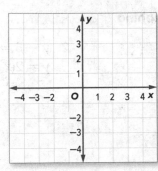

Show your work.

d. _____

e. _____

Example

Tutor

6. A system of equations consists of two lines. One line passes through (2, 3) and (0, 5). The other line passes through (1, 1) and (0, −1). Determine if the system has *no solution, one solution,* or *an infinite number of solutions.*

To compare the two lines, write the equation of each line in slope-intercept form.

Find the slope of each line.

(2, 3) and (0, 5)

$\frac{y_2 - y_1}{x_2 - x_1} = \frac{5 - 3}{0 - 2}$ or −1

(1, 1) and (0, −1)

$\frac{y_2 - y_1}{x_2 - x_1} = \frac{-1 - 1}{0 - 1}$ or 2

Find the *y*-intercept for each line. Then write the equation.

Use the point (0, 5).
The *y*-intercept is 5.

$y = mx + b$
$y = -1x + 5$

Use the point (0, −1).
The *y*-intercept is −1.

$y = mx + b$
$y = 2x - 1$

Since the lines have different slopes and different *y*-intercepts, they intersect in exactly one point.

Check Graph each line on a coordinate plane.

The lines intersect at (2, 3) so there is exactly one solution. ✓

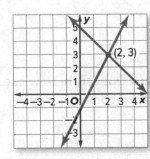

Slopes and Intercepts

When a linear system of equations has:

- different slopes and *y*-intercepts, there is one and only one solution.

- the same slope and different *y*-intercepts, there is no solution.

- the same slope and the same *y*-intercept, there is an infinite number of solutions.

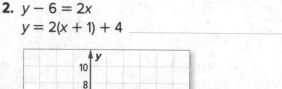

Got it? Do this problem to find out.

f. _____

f. (0, 2), (1, 4) and (0, −1), (1, 1)

Guided Practice

Solve each system of equations by graphing. (Examples 1, 4, and 5)

1. $y = x + 3$
$y = -2x - 3$ _____

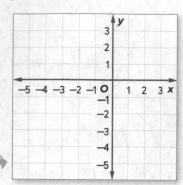

Show your work.

2. $y - 6 = 2x$
$y = 2(x + 1) + 4$ _____

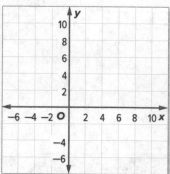

3. The sum of Sally's age plus twice Tomas' age is 12. The difference of Sally's age and Tomas' age is 3. Write and solve a system of equations to find their ages. Interpret the solution. (Examples 2 and 3)

_____
_____
_____

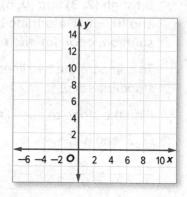

4. A system of equations consists of two lines. One line passes through (−1, 3) and (0, 1). The other line passes through (1, 4) and (0, 2). Determine if the system has *no solution, one solution,* or *an infinite number of solutions.* (Example 6) _____

5. **Building on the Essential Question** How can you use a graph to solve a system of equations?

_____
_____

Rate Yourself!

Are you ready to move on? Shade the section that applies.

YES   ?   NO

For more help, go online to access a Personal Tutor.

Tutor

FOLDABLES Time to update your Foldable!

Solve Systems of Equations Algebraically

Content Standards
8.EE.8, 8.EE.8b, 8.EE.8c

Real-World Link

Jewelry Mary Anne sold 20 necklaces and bracelets at the craft fair. She sold 3 times as many necklaces as bracelets.

Step 1 The bar diagram below represents the situation

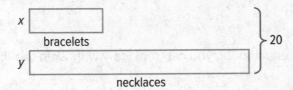

An equation to represent the bar diagram is $x + y = 20$.

Step 2 Mary Anne sold 3 times as many necklaces as bracelets. Divide the necklace bar into sections to represent this.

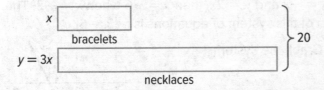

Write an equation using only x to represent the total number of necklaces and bracelets.

_____

Step 3 Solve the equation from Step 2. What does the solution represent? _____

1. How many bracelets and necklaces did Mary Anne sell?

☐ bracelets and ☐ necklaces

Essential Question

WHY are graphs helpful?

Vocabulary

substitution

Common Core State Standards

MP Mathematical Practices
1, 3, 4, 7

Which **MP** Mathematical Practices did you use?
Shade the circle(s) that applies.

① Persevere with Problems   ⑤ Use Math Tools

② Reason Abstractly   ⑥ Attend to Precision

③ Construct an Argument   ⑦ Make Use of Structure

④ Model with Mathematics   ⑧ Use Repeated Reasoning

Solve a System Algebraically

In the previous lesson, you estimated the solution of a system of equations by graphing. **Substitution** is an algebraic model that can be used to find the exact solution of a system of equations.

Example

1. Solve the system of equations algebraically.

$y = x - 3$

$y = 2x$

Since y is equal to $2x$, you can replace y with $2x$ in the first equation.

$y =$	$x - 3$	Write the equation.
$2x =$	$x - 3$	Replace y with 2x.
$-x = -x$		Subtraction Property of Equality
$x = -3$		Simplify.

Since $x = -3$ and $y = 2x$, then $y = -6$ when $x = -3$. The solution of this system of equations is $(-3, -6)$.

Check Graph the system.

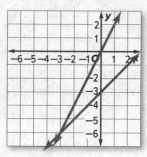

 Got it? Do these problems to find out.

a. _____

b. _____

Solve each system of equations algebraically.

a. $y = x + 4$
  $y = 2$

b. $y = x - 6$
  $y = 3x$

Slope-Intercept and Standard Forms

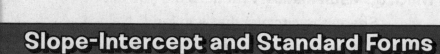

Sometimes one or both equations are written in standard form. When solving a system by substitution, one of the equations should be solved for either x or y.

Example

2. Solve the system of equations algebraically.

$y = 3x + 8$

$8x + 4y = 12$

$8x + 4y =$	12	Write the equation.
$8x + 4(3x + 8) =$	12	Replace y with $3x + 8$.
$8x + 4 \cdot 3x + 4 \cdot 8 =$	12	Distributive Property
$8x + 12x + 32 =$	12	Simplify.
$20x + 32 =$	12	Collect like terms.
$20x + 32 =$	12	
$\underline{-32 = -32}$		Subtraction Property of Equality
$20x =$	-20	Simplify.
$\dfrac{20x}{20} =$	$\dfrac{-20}{20}$	Division Property of Equality
$x =$	-1	Simplify.

Since $x = -1$, replace x with -1 in the equation $y = 3x + 8$ to find the value of y.

$y = 3x + 8$

$y = 3(-1) + 8$ or 5

The solution of this system is $(-1, 5)$.

> **Substitution**
>
> When you replace a variable with an expression, write the expression inside parentheses. This will help you apply the Distributive Property correctly.

Got it? Do these problems to find out.

c. $y = 2x + 1$
   $3x + 4y = 26$

d. $2x + 5y = 44$
   $y = 6x - 4$

Show your work.

c. _____

d. _____

Examples

A total of 75 cookies and cakes were donated for a bake sale to raise money for the football team. There were four times as many cookies donated as cakes.

3. Write a system of equations to represent this situation.

Draw a bar diagram. Then write the system.

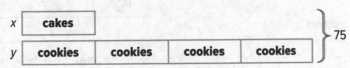

$y = 4x$ There were 4 times as many cookies donated as cakes.

$x + y = 75$ The total number of cakes and cookies is 75.

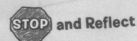
4. Solve the system in Example 3 algebraically. Interpret the solution.

Since y is equal to $4x$, you can replace y with $4x$.

$x + y = 75$	Write the equation.
$x + 4x = 75$	Replace y with $4x$.
$5x = 75$	Simplify.
$\dfrac{5x}{5} = \dfrac{75}{5}$	Division Property of Equality
$x = 15$	Simplify.

Since $x = 15$ and $y = 4x$, then $y = 60$ when $x = 15$. The solution is (15, 60). This means that 15 cakes and 60 cookies were donated.

Got it? Do these problems to find out.

Mr. Thomas cooked 45 hamburgers and hot dogs at a cookout. He cooked twice as many hot dogs as hamburgers.

e. Write a system of equations to represent this situation.

f. Solve the system algebraically. Interpret the solution.

e. _____

f. _____

Show your work.

Guided Practice

Solve each system of equations algebraically. (Examples 1 and 2)

1. $y = x + 7$

$y = 4$ _____

Show your work.

2. $y = x + 5$

$y = 3x$ _____

3. $y = x - 9$

$y = -4x$ _____

4. $x + 3y = 1$

$y = 2x + 5$ _____

5. Seven people went to the movies. The number of adults was one more than the number of children. Write a system of equations that represents the number of adults and children. Solve the system algebraically. Interpret the solution. (Examples 3 and 4)

_____

6. **Building on the Essential Question** How can you solve a system of equations? _____

_____

Independent Practice

Go online for Step-by-Step Solutions
eHelp

Solve each system of equations algebraically. (Examples 1 and 2)

1. $y = x + 5$
$y = 6$

Show your work.

2. $y = x + 12$
$y = -18$

3 $y = x - 10$
$y = -12$

4. $y = x + 15$
$y = 2x$

5. $y = 2x - 3$
$x + y = 18$

6. $y = \frac{1}{4}x$
$x + 4y = 8$

7. $y = x + 12$
$4x + 2y = 27$

8. $10x + 3y = 19$
$y = 2x + 5$

Write and solve a system of equations that represents each situation. Use a bar diagram if needed. Interpret the solution. (Examples 3 and 4)

9. Elaine bought a total of 15 shirts and pairs of pants. She bought 7 more shirts than pants. How many of each did she buy?

_____

_____

10. Together, Preston and Horatio have 49 video games. Horatio has 11 more games than Preston. How many games does each person have?

_____

_____

11 The cost of 8 muffins and 2 quarts of milk is $18. The cost of 3 muffins and 1 quart of milk is $7.50. How much does 1 muffin and 1 quart of milk cost?

_____

_____

12. **MP Multiple Representations** The table shows the rates at which Ajay and Tory are biking along the same trail.

Person	Rate (m/min)
Ajay	200
Tory	250

a. **Algebra** Suppose Ajay began the trail 325 meters ahead of Tory. Write a system of equations to represent the distance y each person will travel after any number of minutes x. _____

b. **Words** Which person was farther along the trail after 5 minutes?

_____

c. **Graphs** Graph the system. Use the graph to determine when Tory will catch up to Ajay.

_____

d. **Algebra** Solve the system of equations algebraically. Interpret your solution. How does your solution compare to your estimate in part **c**?

_____

_____

_____

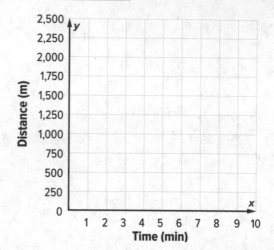

H.O.T. Problems Higher Order Thinking

13. **MP Persevere with Problems** What is the solution to the system $5x + y = 2$ and $y = -5x + 8$? Explain. _____

_____

_____

14. **MP Identify Structure** Describe when it is better to use substitution to solve a system of equations rather than graphing. _____

_____

15. **MP Which One Doesn't Belong?** Circle the system of equations that does not belong with the other three. Explain your reasoning.

$y = 3x - 5$	$y = 5x - 7$	$y = x + 3$	$y = -2x$
$y = -2x$	$y = 2(2x - 3)$	$y = -2x - 3$	$y = -2(3x - 2)$

_____

_____

Extra Practice

Solve each system of equations algebraically.

16. $y = 2x$

$y = x + 1$ (1, 2) _____

$y = x + 1$
$2x = x + 1$
$\underline{-x = -x}$
$x = 1$

Since $x = 1$ and $y = 2x$, $y = 2$.

Homework Help →

17. $y = 4x + 45$

$x = 4y$ _____

18. $y = -2x$

$x = 0$ _____

19. $x + y = -3$

$y = x + 3$ _____

20. $y = x + 4$

$y = 0$ _____

21. $x - y = 6$

$y = -1$ _____

22. The length of the rectangle is 3 meters more than the width. The perimeter is 26 meters. Write and solve a system of equations that represents this situation. What are the dimensions of the rectangle?

_____

_____

23. **MP** **Model with Mathematics** Ms. Corley wants to take her class on a trip to either the nature center or the zoo. The nature center charges $4 per student plus $95 for a 1-hour naturalist program. The zoo charges $9 per student plus $75 for a 1-hour guided tour.

a. Write a system of equations to represent this situation.

_____

b. Solve the system of equations algebraically and by graphing. Interpret the solution.

_____

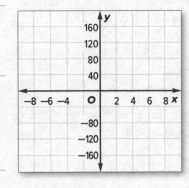

c. Ms. Corley has 22 students in her class. Determine which would cost less, the nature center or the zoo.

_____

24. A veterinarian examined 4 times as many dogs as cats today. In all, he examined 15 dogs and cats. Let *d* represent the number of dogs and *c* represent the number of cats that the veterinarian examined.

c
d
cats
dogs
15

Use the labels to complete the bar diagram that models the situation.

Use the bar diagram to set up and solve a system of equations. How many dogs and cats did the veterinarian examine today?

25. In one volleyball game, Naomi had 3 times as many spikes as Vicki. Together, they had 20 spikes. How many spikes did each player have? Write and solve a system of equations.

CCSS Common Core Spiral Review

Solve. 6.EE.7

26. $p - 12 = 20$ _____

27. $31 = r - 36$ _____

28. $m + 1\frac{3}{8} = 5$ _____

29. $56.9 = 34 + p$ _____

30. $0.97 + a = 2.6$ _____

31. $x - 24 = 73$ _____

32. $t + 5 = 30$ _____

33. $r - 15 = 63$ _____

Inquiry Lab
Analyze Systems of Equations

 HOW can you solve real-world mathematical problems using two linear equations in two variables?

Content Standards
8.EE.8, 8.EE.8a, 8.EE.8b, 8.EE.8c

MP Mathematical Practices
1, 3, 5

A map uses a coordinate grid to show the locations of cities and towns. The map locations for four towns are shown in the table. Suppose Brent travels from Town A to Town B and Maria travels from Town C to Town D. Do Brent's and Maria's routes pass through a common location?

Town	Location
A	(0, 6)
B	(5, 1)
C	(0, 4)
D	(4, 8)

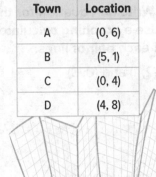

What do you know? _____

_____

What do you need to know? _____

_____

Hands-On Activity

Step 1 Plot and label the points of each town on the coordinate plane shown.

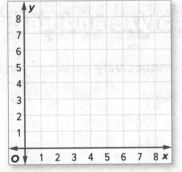

Step 2 Draw a red line segment to represent Brent's route and draw a blue line segment to represent Maria's route.

Step 3 Find the slope of the lines that represent Brent's route and Maria's route.

_____

_____

What do the slopes tell you about the lines? Explain.

_____

_____

Step 4 Where do the two lines intersect? _____

So, Brent's and Maria's routes pass through the common location

(☐ , ☐).

Inquiry Lab Analyze Systems of Equations **251**

Investigate

Refer to the Activity. Work with a partner.

1. Write an equation for the lines that represent Brent's routes and Maria's routes.

   Brent's route: _____   Maria's route: _____

2. Solve the system of equations from Exercise 1 algebraically. _____

MP Use Math Tools Write an equation for the line that passes through each pair of points. Use a graphing calculator to solve the system. Then describe the slope of each pair of lines.

3. (0, −1) and (4, 3); (2, 1) and (0, 3)

   Equations: _____

   Solution: _____

   _____

4. (0, 3) and (3, 9); (0, 2) and (3, 8)

   Equations: _____

   Solution: _____

   _____

Analyze and Reflect

5. **MP Reason Inductively** How can you determine if two lines will intersect

   using the slope? _____

   _____

Create

6. **MP Model with Mathematics** When two lines intersect to form a right angle, the slope of one is the opposite reciprocal of the slope of the other. Write a

   system of equations that form a right angle. _____

   _____

7. **Inquiry** HOW can you solve real-world mathematical problems using two

   linear equations in two variables? _____

   _____

   _____

21ST CENTURY CAREER
in Music

Mastering Engineer

Do you love listening to music? Are you interested in the technical aspects of music-making? If so, a career creating digital masters might be something to think about! A mastering engineer produces digital masters and is responsible for making songs sound better, having the proper spacing between songs, removing extra noises, and assuring all the songs have consistent levels of tone and balance. Having a great-sounding master helps increase radio airplay and sales for recording artists.

College & Career
READINESS

Is This the Career for You?

Are you interested in a career as a mastering engineer? Take some of the following courses in high school.

◆ Algebra
◆ Music Appreciation
◆ Recording Techniques
◆ Sound Engineering

Turn the page to find out how math relates to a career in Music.

ⓜ Mastering the Music

Use the information in the tables to solve each problem.

1. At Engineering Hits, is the relationship between the number of songs and the cost linear? Explain your reasoning. _____

_____

_____

2. Is there a proportional linear relationship between number of songs and cost at Dynamic Mastering? Explain your reasoning.

_____

_____

_____

3. Find the slope of the line represented in the Mastering Mix table. What does the slope represent? _____

_____

4. Is the linear relationship represented in the Mastering Mix table a direct variation? Explain.

_____

_____

5. Write a direct variation equation to represent number of songs x and cost y at Dynamic Mastering. How much does it cost to master 11 songs? _____

6. For 4 or more songs at Engineering Hits, the cost varies directly as the number of songs. How much does it cost to master 6 songs?

Engineering Hits	
Number of Songs	**Cost ($)**
1	100
2	160
3	210
4	250

Dynamic Mastering	
Number of Songs	**Cost ($)**
2	120
4	240
6	360
8	480

Mastering Mix	
Number of Songs	**Cost ($)**
1	125
3	275
5	425
7	575

ⓜ Career Project

It's time to update your career portfolio! Find the name of the mastering engineer on one of your CDs. Use the Internet or another source to write a short biography of this engineer. Include a list of other artists whose songs he or she has mastered.

_____

_____

_____

_____

Do you think you would enjoy a career as a mastering engineer? Why or why not?

_____

_____

_____

_____

Chapter Review ✓ Check

Vocabulary Check

Complete the crossword puzzle using the vocabulary list at the beginning of the chapter.

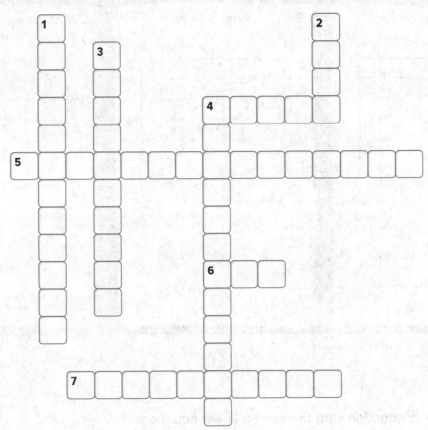

Across

4. a value to describe the steepness of a straight line

5. a relationship in which the ratio of two variable quantities is constant

6. the horizontal change between the same two points

7. the x-coordinate of the point where the graph crosses the x-axis

Down

1. an algebraic model used to find the exact solution of a system of equations

2. the vertical change between any two points

3. the y-coordinate of the point where the line crosses the y-axis

4. when an equation is written in the form $Ax + By = C$

Use Your FOLDABLES

Use your Foldable to help review the chapter.

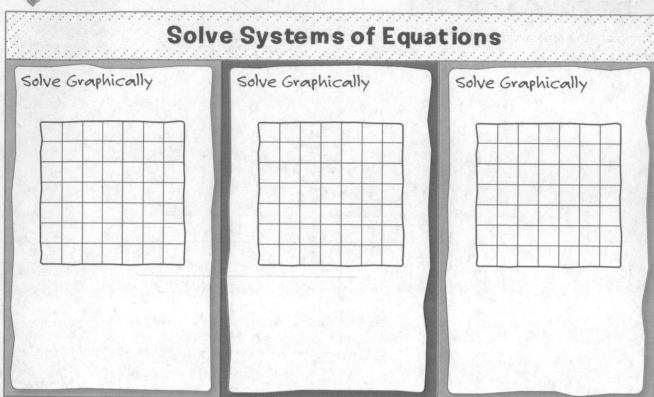

Tape here

Solve Systems of Equations

Solve Graphically

Solve Graphically

Solve Graphically

Got it?

Match each set of information with the correct linear equation.

1. line that passes through (2, 0) and (0, 1)

   a. $y = 0.5x$

2. line with a slope of 0.5 and a y-intercept of 1

   b. $x = 5$

3. line that passes through (4, 2) and the origin

   c. $y = 0.5x + 1$

4. line that has a slope of 0 and passes through (5, 4)

   d. $y = -0.5x + 1$

5. line that has an undefined slope and passes through (5, 4)

   e. $y = 4$

Power Up! Performance Task

Play for a Prize

Antoine likes to go to the arcade. Tickets are awarded in different ways for his favorite games.

Game #1	Game #2
two tickets per game PLUS one ticket for every sixty points scored	one ticket for every forty points scored

- The ticket dispenser awards the customer tickets based on the point total.
- For each game, x represents the number of points scored and y represents the number of tickets awarded.

Write your answers on another piece of paper. Show all of your work to receive full credit.

Part A
Write an equation to model the number of tickets earned for each game. Interpret the slope and y-intercept for each situation.

Part B
Is either relationship proportional? Explain your reasoning.

Part C
Solve the system of equations algebraically. Interpret the solution.

Part D
Suppose Antoine needs 10 more tickets to win a prize. Which game should he play? Explain your reasoning.

Reflect

 Answering the Essential Question

Use what you learned about graphs to complete the graphic organizer. List three ways in which graphs are helpful. Then give an example for each way.

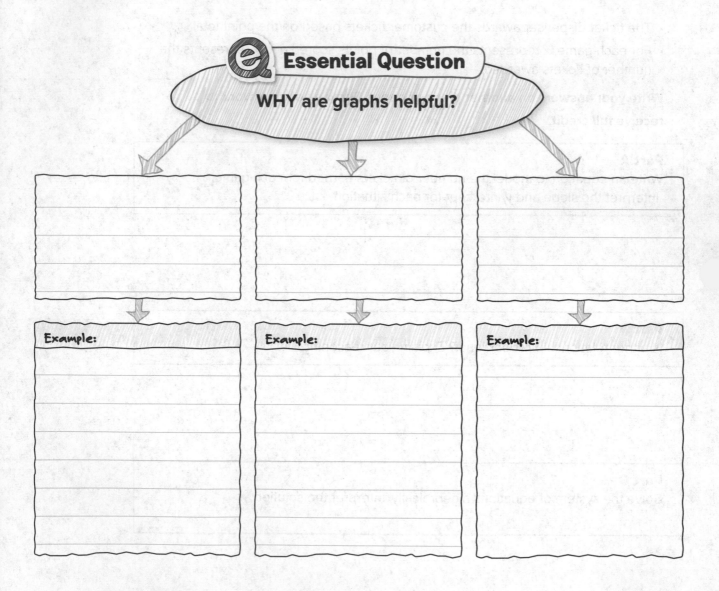

Essential Question

WHY are graphs helpful?

Example:

Example:

Example:

 Answer the Essential Question. WHY are graphs helpful?

UNIT PROJECT

Watch ▶

Web Design 101 When designing a good Web page, there are many details to consider in order to make your Web page stand out. In this project, you will:

- **Collaborate** with your classmates as you research an animal and design a Web page.
- **Share** the results of your research in a creative way
- **ℯ Reflect** on how you communicate mathematical ideas effectively.

By the end of this Project, you will be ready to design a live Web page about your favorite animal!

Collaborate

⏻ Go Online Work with your group to research and complete each activity. You will use your results in the Share section on the following page.

1. Choose your favorite animal. Research information about that animal, such as the population over the past 10 years, the kinds of food it eats, sleeping habits, its average lifespan, average size, and average speed. Present this information using tables and graphs.

2. Use the distance formula, distance = rate × time, to write an equation that represents the distance your animal can travel at its average speed. Find the average speed of two other animals and write equations using the distance formula. Graph all three equations on the same coordinate plane. Then describe the graphs.

3. Research the elements needed to make a good Web page. Then make a sketch of your own Web page about the favorite animal that you selected in Exercise 1. Be sure to include tables, equations, graphs, and photos.

4. Find another animal in the same animal kingdom as your favorite animal. On the sketch of your Web page, include a link to this other animal and an equation that describes one of its characteristics.

5. Research the cost of taking a Web design class. Write an equation that represents the time it will take you to save enough money for the class. Share this equation as you write a few paragraphs that explain your plan on how to save enough money.

With your group, decide on a way to share what you have learned about your animal and Web pages. Some suggestions are listed below, but you can also think of other creative ways to present your information. Remember to show how you used mathematics to complete each of the activities in this project!

- If possible, use Web page creation software to turn your design into a live Web page.
- Imagine you are going to be interviewed by a reporter about your work on this project. Write down what will be discussed in the interview. You may wish to actually record an interview.

Check out the note on the right to connect this project with other subjects.

connect with Economics

Business Literacy Research Web design jobs in your area. Find out the following:

- What type of education is required?
- What skills should a Web designer possess?

Reflect

On Your Own

6. **ⓔ Answer the Essential Question** HOW can you communicate mathematical ideas effectively?

   a. How did you use what you learned in the Equations in One Variable chapter to communicate mathematical ideas effectively in this project?

   _____

   _____

   _____

   b. How did you use what you learned in the Equations in Two Variables chapter to communicate mathematical ideas effectively in this project?

   _____

   _____

   _____

UNIT 3

CCSS Functions

Essential Question

HOW can you find and use patterns to model real-world situations?

Chapter 4
Functions

Functions can be represented using equations, graphs, tables, and verbal descriptions. In this chapter, you will use functions to model linear relationships. You will also investigate nonlinear functions.

Unit Project Preview

Watch

Green Thumb Do you have a green thumb for gardening? A community garden is a great way to meet your neighbors, beautify your neighborhood, and strengthen the sense of community.

There are many types of community gardens. Food pantry gardens donate the produce they grow to local food pantries. School gardens help to educate students in science and math, while entrepreneurial gardens generate income by selling the produce.

At the end of Chapter 4, you'll complete a project to discover the costs involved in creating a community garden. But for now, it's time to do an activity in your book. Complete the table shown by estimating the cost of selling various vegetables and fruits.

My Garden	
Item	**Cost Per Item or Per Pound**
Carrots	
Cucumbers	
Peas	
Strawberries	
Tomatoes	

Chapter 4
Functions

Essential Question

HOW can we model relationships between quantities?

Common Core State Standards

Content Standards
8.F.1, 8.F.2, 8.F.3, 8.F.4, 8.F.5

 Mathematical Practices
1, 2, 3, 4, 5, 7

Math in the Real World

Free Throws Each year the middle school sponsors a free throw contest. Ms. Meyer's class has to find the height of the basketball after a given amount of time. The equation $y = -16x^2 + 20x + 5$ can be used to find the height y in feet of the ball after x seconds. Use the graph below to find the height of the basketball after 0.5 second.

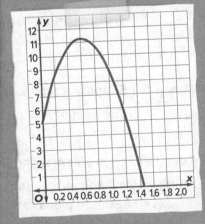

Study Organizer

 Cut out the Foldable on page FL9 of this book.

 Place your Foldable on page 358.

 Use the Foldable throughout this chapter to help you learn about functions.

What Tools Do You Need?

Vocabulary

continuous data

dependent variable

discrete data

domain

function

function table

independent variable

linear equation

linear function

nonlinear function

quadratic function

qualitative graphs

range

relation

Study Skill: Reading Math

One way to make a word problem easier to understand is to rewrite it using fewer words.

Step 1 | **Read the problem and identify the important words and numbers.**

There are a great deal of cell phone plans available for students. With Janelle's plan, she pays **$15 per month** for 200 minutes, plus $0.10 per minute once she talks for more than 200 minutes. Suppose Janelle can spend $20 each month for her cell phone. How many more minutes can she talk?

Step 2 | **Simplify the problem. Keep all of the important words and numbers, but use fewer of them.**

The total monthly cost is **$15** for 200 minutes, plus $0.10 times the number of minutes over 200. How many minutes can she talk for $20?

Step 3 | **Simplify it again. Use a variable for the unknown.**

The cost of *m* minutes at $0.10 per minute plus **$15** is $20.

Rewrite each problem using the method above.

1. Akira is saving money to buy a scooter that costs $125. He has already saved $80 and plans to save an additional $5 each week. In how many weeks will he have enough money for the scooter?

2. Joaquin wants to buy some DVDs that are each on sale for $10 plus a CD that costs $15. How many DVDs can he buy if he has $75 to spend?

What Do You Already Know?

Place a checkmark below the face that expresses how much you know about each concept. Then scan the chapter to find a definition or example of it.

 I know it.   I've heard of it.   I have no clue.

Integers				
Concept	😟	😐	😊	Definition or Example
domain and range				
functions				
independent and dependent variables				
multiple representations of linear functions				
nonlinear functions				
relations				

When Will You Use This?

Here are a few examples of how functions are used in the real world.

Activity 1 Use the Internet to research the cost of printing and shipping photos from two different photo printing services. Which company offers the better deal?

_____

Activity 2 Go online at **connectED.mcgraw-hill.com** to read the graphic novel **Picture This**. How much will it cost for Brian to print and ship his pictures?

_____

Brian and Dion in
Picture This
CLIK!

Are You Ready?

Try the Quick Check below.
Or, take the Online Readiness Quiz.

Common Core Review 6.NS.6, 7.NS.3

Example 1

Name the ordered pair for point Q.

Start at the origin. Move right along the x-axis until you reach 1.5. Then move up until you reach the y-coordinate, 2. Point Q is located at (1.5, 2).

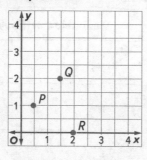

Example 2

Evaluate 6x + 1 if x = −4.

$6x + 1 = 6(-4) + 1$ Replace x with −4.

$= -24 + 1$ Multiply 6 by −4.

$= -23$ Add.

Quick Check

Coordinate Graphing Name the ordered pair for each point.

1. R _____

2. S _____

3. T _____

4. U _____

5. V _____

6. W _____

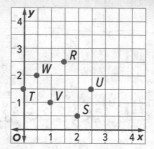

Evaluate Expressions Evaluate each expression if x = −6.

7. $3x$ _____

8. $4x + 9$ _____

9. $\dfrac{x}{2}$ _____

10. $\dfrac{3x}{9}$ _____

11. The weekly profit of a certain company is 48x − 875, where x represents the number of units sold. Find the weekly profit if the company sells 37 units. _____

 **Which problems did you answer correctly in the Quick Check? Shade those exercise numbers below.**

① ② ③ ④ ⑤ ⑥ ⑦ ⑧ ⑨ ⑩ ⑪

Representing Relationships

 Real-World Link Watch

Space To achieve orbit, the space shuttle must travel at a rate of about 5 miles per second. The table shows the total distance *d* that the craft covers in certain periods of time *t*.

Time *t* (seconds)	Distance *d* (miles)
1	5
2	10
3	15
4	20
5	25

1. Write an algebraic expression for the distance in miles for any number of seconds *t*. _____

2. Describe the relationship in words.

_____

3. Graph the ordered pairs. Describe the shape of the graph.

_____

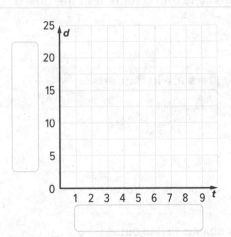

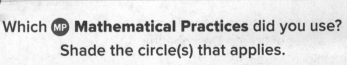

 Essential Question

HOW can we model relationships between quantities?

Vocab
$\frac{a}{b_c}$ **Vocabulary**

linear equation

CCSS **Common Core State Standards**

Content Standards
8.F.4

MP Mathematical Practices
1, 3, 4, 5

Which MP **Mathematical Practices** did you use?
Shade the circle(s) that applies.

① Persevere with Problems

② Reason Abstractly

③ Construct an Argument

④ Model with Mathematics

⑤ Use Math Tools

⑥ Attend to Precision

⑦ Make Use of Structure

⑧ Use Repeated Reasoning

Tables, Graphs, and Equations

Recall that an equation is a mathematical sentence stating that two quantities are equal. A **linear equation** is an equation with a graph that is a straight line. Some equations contain more than one variable.

 Real World

Examples

 Tutor

The table shows the number of liters in quarts of liquid.

1. Write an equation to find the number of liters in any number of quarts. Describe the relationship in words.

The rate of change is the rate that describes how one quantity changes in relation to another quantity. The rate of change of quarts to liters is $\frac{1.9 - 0.95}{2 - 1} = \frac{0.95}{1}$ or 0.95 liter in every quart.

Quarts, q	Liters, ℓ
1	0.95
2	1.9
3	2.85
4	3.8
5	4.75

+0.95
+0.95
+0.95
+0.95

Let ℓ represent the liters and q represent the quarts. The equation is $\ell = 0.95q$.

2. About how many liters are in 8 quarts?

$\ell = 0.95q$ Write the equation.

$\ell = 0.95(8)$ Replace q with 8.

$\ell = 7.6$ Multiply.

There are about 7.6 liters in 8 quarts.

> **Variables**
>
> Recall you can use any letter to represent the independent and dependent variables. If you graph the equation, label your axes using those letters.

Show your work.

Got it? Do these problems to find out.

The total cost of tickets to the school play is shown in the table.

a. Write an equation to find the total cost of any number of tickets. Describe the relationship in words.

b. Use the equation to find the cost of 15 tickets.

Number of Tickets, t	Total Cost (\$), c
1	4.50
2	9.00
3	13.50
4	18.00

a. _____

b. _____

Examples

The total distance Marlon ran in one week is shown in the graph.

3. Write an equation to find the number of miles ran y after any number of days x.

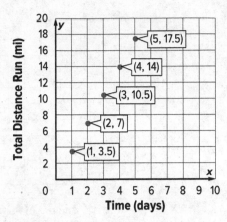

Find the rate of change or the slope of the line.

Step 1 $m = \dfrac{y_2 - y_1}{x_2 - x_1}$ Definition of slope

$m = \dfrac{14 - 7}{4 - 2}$ $(x_1, y_1) = (2, 7); (x_2, y_2) = (4, 14)$

$m = \dfrac{7}{2}$ or 3.5   Simplify.

Step 2 To find the y-intercept, use the slope and the coordinates of a point to write the equation of the line in slope-intercept form.

$y = mx + b$ Slope-intercept form

$y = 3.5x + b$ Replace m with the slope, 3.5.

$7 = 3.5(2) + b$ Use the point (2, 7). $x = 2, y = 7$

$0 = b$ Solve for b.

The slope is 3.5 and the y-intercept is 0. So, the equation of the line is $y = 3.5x + 0$ or $y = 3.5x$.

- -

4. How many miles will Marlon run after 2 weeks?

$y = 3.5x$ Write the equation.

$y = 3.5(14)$ There are 14 days in 2 weeks. Replace x with 14.

$y = 49$ Multiply.

Marlon will run 49 miles in 2 weeks.

The number of trees saved by recycling paper is shown.

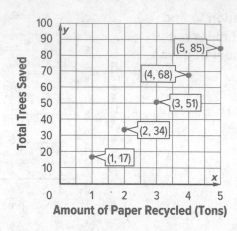

c. _____

d. _____

c. Write an equation to find the total number of trees y that can be saved for any number of tons of paper x.

d. Use the equation to find how many trees could be saved if 500 tons of paper are recycled.

Key Concept

Multiple Representations of Linear Equations

Words

Distance traveled is equal to 12 miles per second times the number of seconds.

Equation

$$d = 12s$$

Table

Time (seconds)	Distance (miles)
1	12
2	24
3	36
4	48
5	60

Graph

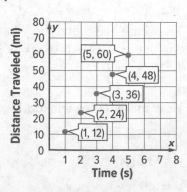

Words, equations, tables, and graphs can be used to represent linear relationships.

Examples

Chloe competes in jump rope competitions. Her average rate is 225 jumps per minute.

5. Write an equation to find the number of jumps in any number of minutes.

Let j represent the number of jumps and m represent the minutes.

The equation is $j = 225m$.

6. Make a table to find the number of jumps in 1, 2, 3, 4, or 5 minutes. Then graph the ordered pairs.

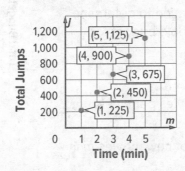

m	225m	j
1	225(1)	225
2	225(2)	450
3	225(3)	675
4	225(4)	900
5	225(5)	1,125

Got it? Do these problems to find out.

Financial Literacy Paul earns $25 for grooming a dog plus $18.50 per day for boarding the same dog.

e. Write an equation to find the amount of money Paul earned m for grooming a dog once and boarding it for any number of days d.

f. Make a table to find his earnings for 5, 6, 7, or 8 days. Then graph the ordered pairs.

 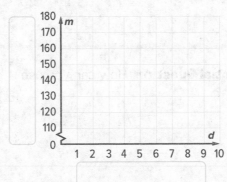

e. _____

Show your work.

1. The table shows the total number of text messages that Brad sent over 4 days. (Examples 1 and 2)

Number of Days, d	1	2	3	4
Total Messages, m	50	100	150	200

   a. Write an equation to find the total number of messages sent in any number of days. Describe the relationship in words.

   _____

   b. Use the equation to find how many text messages

   Brad would send in 30 days. _____

2. **Financial Literacy** The graph shows the amount of money the Rockwell family budgets for food each month. Write an equation to find the total amount of money c budgeted in any number of months m. Use the equation to determine how much money the Rockwell family should budget for

   12 months. (Examples 3 and 4) _____

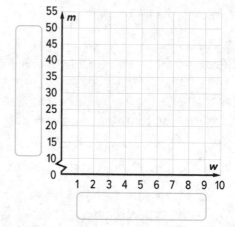

3. A store receives an average of 7 new movies per week. (Examples 5 and 6)

   a. Write an equation to find the number of new movies m in

   any number of weeks w. _____

   b. Make a table to find the number of new movies received in 4, 5, 6, or 7 weeks. Then graph the ordered pairs.

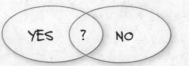

4. **Building on the Essential Question** How can you use a graph to write an equation?

   _____

   _____

   _____

Independent Practice

Go online for Step-by-Step Solutions

1 The number of baskets a company produces each day is shown in the table. (Examples 1 and 2)

 Show your work.

a. Write an equation to find the total number of baskets crafted in any number of days. Describe the relationship in words.

_____

b. Use the equation to determine how many baskets the company makes in one non-leap year. _____

Number of Days, d	Total Baskets, b
1	45
2	90
3	135
4	180

2. A type of dragonfly is the fastest insect. The graph shows how far the dragonfly can travel. (Examples 3 and 4)

a. Write an equation to find how far the dragonfly can travel d in any number of seconds s. _____

b. Use the equation to determine how far the dragonfly can travel in one minute. _____

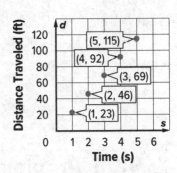

3 A library charges a late return fee of $3.50 plus $0.15 per day that a book is returned late. (Examples 5 and 6)

a. Write an equation to find the total late fee f for any number of days late d. _____

b. Make a table to find the total fee if a book is 10, 15, 20, or 25 days late. Then graph the ordered pairs.

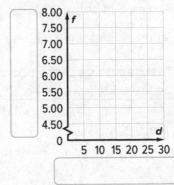

4. **MP Multiple Representations** The two fastest times for swimming the English Channel belong to Petar Stoychev and Yvetta Hlaváčová. Petar's average speed was 265 feet per minute. Yvetta's average speed was 249 feet per minute.

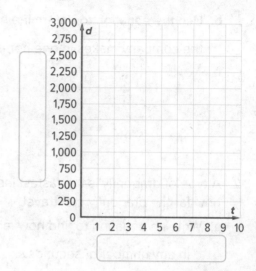

Time (min)	Petar	Yvetta
1	265	249

a. **Tables** Complete the table of ordered pairs in which the x-coordinate represents the time and the y-coordinate represents the total distance swum in 1, 2, 3, 4, and 5 minutes.

b. **Graphs** Graph each set of ordered pairs on the coordinate plane.

c. **Algebra** Write an equation for each swimmer to find the number of feet swam d in any number

of minutes t.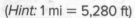

d. **Numbers** If Petar Stoychev swam the Channel in 6 hours, 57 minutes, and 50 seconds, approximately how wide in miles is the English Channel?

(*Hint:* 1 mi = 5,280 ft)

H.O.T. Problems Higher Order Thinking

5. **MP Model with Mathematics** Write an equation with two variables that represents a real-world situation. _____

_____

6. **MP Persevere with Problems** The table shows the areas of circles with radii from 1 through 3 feet.

Radius (ft), r	1	2	3
Area (ft²), A	π	4π	9π

Recall that π has a value of about 3. Write an equation in two variables to

represent the relation in the table. _____

7. **MP Model with Mathematics** Write about a real-world situation that can be represented by the equation $y = 4x$.

_____

_____

_____

Extra Practice

8. The table shows the number of square inches per square foot.

a. Write an equation to find the number of square inches *i* in any number of square feet *f*. Describe the relationship in words.

i = 144f; There are 144 square inches in every square foot.

⟶ The rate of change is $\frac{288 - 144}{2 - 1}$ or 144 square inches for every square foot. So the equation is i = 144f.

Square Feet, f	Square Inches, i
1	144
2	288
3	432
4	576

b. Use the equation to determine how many square inches are in 15 square feet. 2,160 square inches

Use the equation i = 144f.

$$i = 144(15)$$
$$i = 2,160$$

9. Financial Literacy Kara is saving money for a school trip. The graph shows how much money she has saved over 4 weeks.

a. Write an equation to find how much money *d* Kara can save over *w* weeks. _____

b. Use the equation to determine how much money Kara can save in 24 weeks. _____

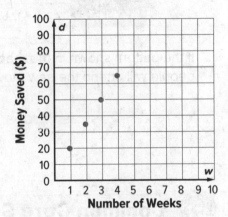

10. **MP** **Use Math Tools** Cornett Cable charges $32.50 a month for basic cable television. Each premium channel selected costs an additional $4.95 per month.

a. Write an equation to find the total monthly cost *c* for any number of premium channels *p*. _____

b. Make a table to show the monthly cost for 0, 1, 2, 3, and 4 premium channels. Then graph the ordered pairs.

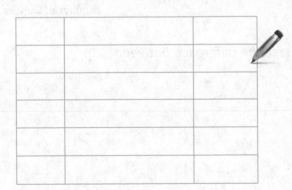

11. The graph represents the total cost to send text messages. Based on the graph, which of the following costs are correct? Select all that apply.

☐ 32 text messages cost $4.80

☐ 50 text messages cost $7.50

☐ 60 text messages cost $9.50

☐ 70 text messages cost $10.50

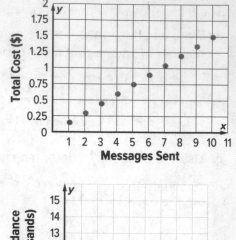

12. The table shows the number of people who attended a new movie over the course of a week. Graph the relationship on the coordinate plane.

Day	1	3	5	7
Attendance	12,200	12,600	13,000	13,400

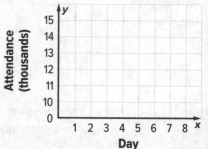

If the pattern shown in the graph continues, how many people will attend the new movie on the 8th day?

Common Core Spiral Review

13. The graph at the right shows the approximate number of grams in one ounce. **6.EE.2**

a. Write an algebraic equation to represent the data in the graph.

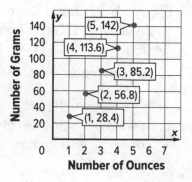

b. Use the expression to find the number of grams in 150 ounces.

14. Write an algebraic expression to represent the phrase *the difference between two times b and eleven.* **6.EE.2**

Copyright © McGraw-Hill Education   Ciaran Griffin/Stockbyte/Getty Images

Vocabulary Start-Up

Complete the graphic organizer of the coordinate plane below.

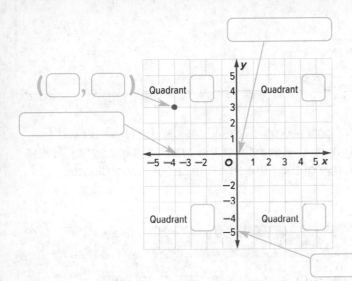

(☐ , ☐)

Quadrant ☐   Quadrant ☐

Quadrant ☐   Quadrant ☐

Identify the x-coordinate in the point (−5, −7).

_____

Real-World Link

How do maps use the coordinate plane for locating towns?

_____

_____

_____

Which MP Mathematical Practices did you use?
Shade the circle(s) that applies.

① Persevere with Problems   ⑤ Use Math Tools

② Reason Abstractly   ⑥ Attend to Precision

③ Construct an Argument   ⑦ Make Use of Structure

④ Model with Mathematics   ⑧ Use Repeated Reasoning

Essential Question

HOW can we model relationships between quantities?

Vocabulary

relation
domain
range

Common Core State Standards

Content Standards
Preparation for 8.F.1

MP **Mathematical Practices**
1, 3, 4, 7

Nowheresville, Population 1

Relations

| Ordered Pairs | Table | Graph |

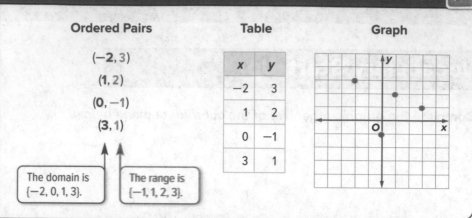

(−2, 3)

(1, 2)

(0, −1)

(3, 1)

x	y
−2	3
1	2
0	−1
3	1

The domain is {−2, 0, 1, 3}.

The range is {−1, 1, 2, 3}.

A **relation** is any set of ordered pairs. Relations can be represented as a table and as a graph. The **domain** of the relation is the set of *x*-coordinates. The **range** of the relation is the set of *y*-coordinates.

Example

Tutor

1. **Express the relation {(2, 6), (−4, −8), (−3, 6), (0, −4)} as a table and a graph. Then state the domain and range.**

Place the ordered pairs in a table with *x*-coordinates in the first column and the *y*-coordinates in the second column.

x	y
2	6
−4	−8
−3	6
0	−4

Graph the ordered pairs on a coordinate plane.

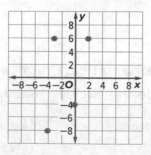

The domain is {−4, −3, 0, 2}. The range is {−8, −4, 6}.

Domain and Range

If a term in the domain or range appears more than once, only write it one time. In Example 1, the value 6 appears twice in the range.

Got it? Do this problem to find out.

a. Express the relation {(−5, 2), (3, −1), (6, 2), (1, 7)} as a table and a graph. Then state the domain and range.

x	y

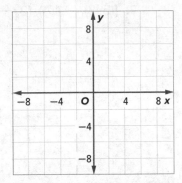

a. _____

Show your work.

Example

Tutor

2. It costs $3 per hour to park at the Wild Wood Amusement Park.

a. Make a table of ordered pairs in which the x-coordinate represents the hours and the y-coordinate represents the total cost for 3, 4, 5, and 6 hours.

x	y
3	9
4	12
5	15
6	18

b. Graph the ordered pairs.

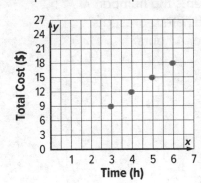

Total Cost ($) vs Time (h)

Got it? Do these problems to find out.

A movie rental store charges $3.95 per movie rental.

b. Make a table of ordered pairs in which the x-coordinate represents the number of movies rented and the y-coordinate represents the total cost for 1, 2, 3, or 4 movies.

c. Graph the ordered pairs.

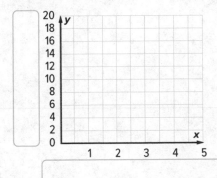

b. _____

x	y

Express each relation as a table and a graph. Then state the domain and range. (Example 1)

1. {(−4, 3), (2, 1), (0, 3), (−3, −2)}

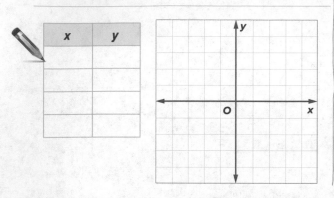

x	y

2. {(5, 3), (−4, 1), (2, −5), (3, −4)}

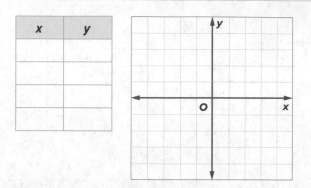

x	y

3. At a vacation resort, you can rent a personal watercraft for $20 per hour. (Example 2)

a. Make a table of ordered pairs in which the *x*-coordinate represents the number of hours and the *y*-coordinate represents the total cost for 1, 2, 3, or 4 hours.

x	y

b. Graph the ordered pairs.

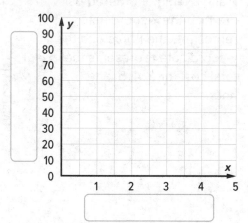

4. 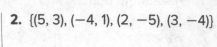 **Building on the Essential Question** How do tables and graphs represent relations?

_____

_____

_____

Rate Yourself!

How confident are you about relations? Check the box that applies.

For more help, go online to access a Personal Tutor.

Independent Practice

Go online for Step-by-Step Solutions eHelp

Express each relation as a table and a graph. Then state the domain and range. (Example 1)

🏠**1** {(8, 5), (−6, −9), (2, 5), (0, −8)}

2. $\left\{\left(2\frac{1}{2}, -1\frac{1}{2}\right), \left(2, \frac{1}{2}\right), \left(-1, 2\frac{1}{2}\right), \left(-1, -1\frac{1}{2}\right)\right\}$

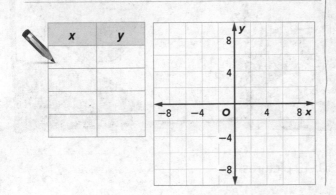

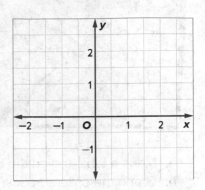

Copy and Solve Draw the table and graph on a separate sheet of paper. A company can manufacture 825 small cars per day. (Example 2)

3. Make a table of ordered pairs in which the *x*-coordinate represents the number of days and the *y*-coordinate represents the total number of cars produced in 1, 2, 3, 4, and 5 days.

4. Graph the ordered pairs.

5 🅼🅿 **Multiple Representations** Refer to the table at the right.

   a. **Words** Describe the pattern, if any, in the table. _____

   _____

   b. **Numbers** Write the ordered pairs (*x*, *y*).

   _____

   c. **Graphs** Graph the ordered pairs on a coordinate plane.

   d. **Words** Describe the graph. How is it different from the other real-world graphs in this lesson?

   _____

   _____

x	y
1	1
2	4
3	9
4	16
5	25

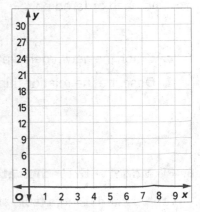

6. **MP Model with Mathematics** Refer to the graphic novel frame below for Exercises a–c. Show your work on a separate sheet of paper.

a. Make a table to find the cost to print 10, 20, 30, 40 pictures.

b. Graph the ordered pairs.

c. How much would it cost for Brian to print and ship 75 pictures? 100?

H.O.T. Problems Higher Order Thinking

7. **MP Persevere with Problems** Refer to the table at the right.

a. Graph the ordered pairs.

b. Reverse the y-coordinates and x-coordinates in each ordered pair.

_____

x	y
0	1
1	3
2	5
3	7

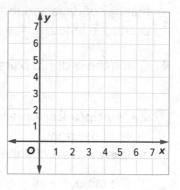

c. Graph the new ordered pairs on the same coordinate plane in part a.

d. Describe the relationship between the two sets of ordered pairs.

_____

_____

8. **MP Model with Mathematics** Describe a real-world situation that can be

represented using a table and a graph. _____

_____

9. **MP Find the Error** Morgan says that the domain of the relation {(2, 3), (−4, 2),

(0, −4), (1, 5)} is {−4, 2, 3, 5}. Find her mistake and correct it. _____

_____

Extra Practice

Express each relation as a table and a graph. Then state the domain and range.

10. {(9, 4), (5, −7), (−3, −4), (−8, 7)}

D: {−8, −3, 5, 9}; R: {−7, −4, 4, 7}

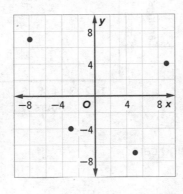

x	y
9	4
5	−7
−3	−4
−8	7

11. {(−1.5, 3.5), (2.5, −1.5), (3, −1), (−1.5, −3.5)}

x	y

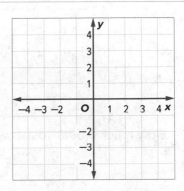

12. A candy company produces 30 boxes of candy per hour.

a. Make a table of ordered pairs in which the *x*-coordinate represents the number of hours and the *y*-coordinate represent the number of boxes of candy in 5, 10, 15, and 20 hours.

x	y

b. Graph the ordered pairs.

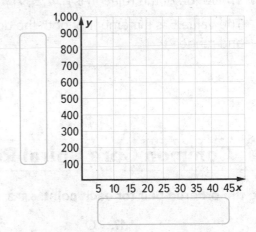

13. **MP Identify Structure** Graph the points in the table on a coordinate plane. Label the points *A*, *B*, and *C*. What are the coordinates of point *D* if points *A*, *B*, *C*, and *D* form a square?

x	y
1	4
−4	4
−4	−1

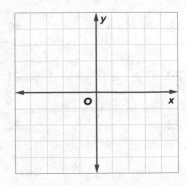

14. Josiah earns $7 an hour for washing cars as a summer job. Complete the table of ordered pairs to show his total earning for several hours. Then express the relation as a graph.

Hours Worked	Total Earned
1	
2	
3	
4	
5	

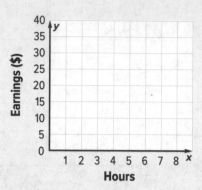

How much would Josiah earn for 12 hours of washing cars?

15. Determine if each statement about the relation {(3, 7), (5, 1), (6, 4), (2, 5)} is true or false.

a. The domain of the relation is {2, 3, 5, 6}. ☐ True ☐ False

b. The range of the relation is {1, 4, 5, 7}. ☐ True ☐ False

c. The value 5 is a member of both the domain and range. ☐ True ☐ False

ⓒⓒⓢⓢ **Common Core Spiral Review**

Name the ordered pair for each point. 6.RP.3

16. P _____

17. Q _____

18. R _____

19. S _____

20. T _____

21. U _____

22. V _____

23. W _____

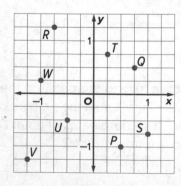

Inquiry Lab
Relations and Functions

 Inquiry HOW can I determine if a relation is a function?

 CCSS Content Standards 8.F.1

MP Mathematical Practices 1, 3, 4

Mrs. Heinl asked three members of her class their favorite color. The mapping diagrams below show some possible results.

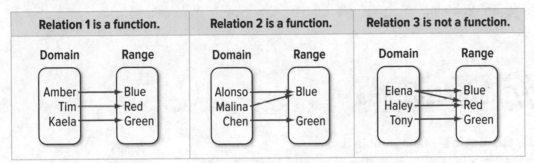

| Relation 1 is a function. | Relation 2 is a function. | Relation 3 is not a function. |

Domain — Range

Relation 1: Amber → Blue, Tim → Red, Kaela → Green

Relation 2: Alonso → Blue, Malina → Blue, Chen → Green

Relation 3: Elena → Blue, Elena → Red, Haley → Red, Tony → Green

A *function* is a special relation in which each member of the domain is paired with *exactly* one member in the range. In the mapping above, Relation 3 is *not* a function because Elena chose two favorite colors, blue and red.

Hands-On Activity

Mr. Morgan asked his students how many pets they have. Some of the student responses are shown in the table.

Student Number	1	3	6
Number of Pets	2	5	7

Complete the mapping diagram shown.

Is the relation a function? Explain.

Domain   Range

1
3
6

_____

_____

Suppose Student 8 has 2 pets. Make a mapping diagram of this situation. Is this relation a function? Explain.

Domain   Range

1   2
3   5
6   7
8

_____

_____

Investigate

1. **MP Model with Mathematics** Students were asked about the number of cell phone minutes they use. Some of the responses are shown in the table. Make a mapping diagram for the relation.

Student	Sarah	Max	Jacob	Rebekah
Number of Minutes	275	220	350	275

Is this relation a function? Explain. _____

_____

Domain   **Range**

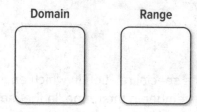

Analyze and Reflect

2. **MP Model with Mathematics** Make a table and a mapping diagram for the relation {(0, −2), (1, −2), (1, 3), (1, 8)}.

Domain			
Range			

Is this relation a function? Explain. _____

_____

Domain   **Range**

Create

3. **MP Use Math Tools** Think of a real-world situation that is not a function. Complete the table and mapping diagram for your situation.

Explain why your situation is not a function. _____

_____

4. **inquiry** HOW can I determine if a relation is a function?

_____

_____

Vocabulary Start-Up

A **function** is a relation in which every member of the domain (input value) is paired with exactly one member of the range (output value). An example of a function is $m = 20n$, where m represents the amount of money earned and n represents the number of lawns mowed. In this example, n is the *independent variable* and m is the *dependent variable*.

> Independent Variable
> What I think it means
> _____
> _____

> Dependent Variable
> What I think it means
> _____
> _____

For each situation, determine which unknown is the dependent variable and which one is the independent variable.

Independent Variable	Equation	Dependent Variable
number of downloads	The equation $c = 0.99n$ represents the total cost c for n music downloads.	cost
	The equation $d = 4.5h$ represents the number of miles d Amber can run in h hours.	
	The equation $s = g + 3$ represents the final score of the game s after g goals in the final period.	

Which **MP** Mathematical Practices did you use?
Shade the circle(s) that applies.

① Persevere with Problems

② Reason Abstractly

③ Construct an Argument

④ Model with Mathematics

⑤ Use Math Tools

⑥ Attend to Precision

⑦ Make Use of Structure

⑧ Use Repeated Reasoning

Essential Question

HOW can we model relationships between quantities?

Vocabulary

function
function table
independent variable
dependent variable

Common Core State Standards

Content Standards
8.F.1, 8.F.4

MP Mathematical Practices
1, 2, 3, 4

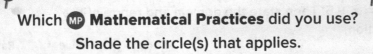

Functions

To find the value of a function for a certain number, substitute the number for the variable x.

$$f(x) = 15x$$

$f(x)$ is read *the function of x*, or *f of x*.
It is the *output* or range.

The *input x*
is any real number. It is the domain.

Example

Tutor

1. Find $f(-3)$ if $f(x) = 2x + 1$.

$f(x) = 2x + 1$ Write the function.

$f(-3) = 2(-3) + 1$ Substitute -3 for x into the function rule.

$f(-3) = -6 + 1$ or -5 Simplify.

So, $f(-3) = -5$.

Got it? Do these problems to find out.

Show your work.

a. _____

b. _____

Find each function value.

    a. $f(2)$ if $f(x) = x - 4$ **b.** $f(11)$ if $f(x) = \frac{1}{2}x + 5$

Function Tables

Watch

You can organize the input, rule, and output into a **function table**. The variable for the domain is called the **independent variable** because it can be any number. The variable for the range is called the **dependent variable** because it depends on the domain.

Example

Tutor

2. Choose four values for x to make a function table for $f(x) = x + 5$. Then state the domain and range of the function.

Substitute each domain value x into the function rule. Then simplify to find the range value.

The domain is $\{-2, -1, 0, 1\}$.
The range is $\{3, 4, 5, 6\}$.

Domain	Rule	Range
x	$f(x) = x + 5$	$f(x)$
-2	$-2 + 5$	3
-1	$-1 + 5$	4
0	$0 + 5$	5
1	$1 + 5$	6

Got it? **Do this problem to find out.**

   c. Choose four values for *x* to complete the function table for the function $f(x) = x - 7$. Then state the domain and range of the function.

x	$f(x) = x - 7$	$f(x)$

c. _____

Tutor

Examples

There are approximately 770 peanuts in a jar of peanut butter. The total number of peanuts $p(j)$ is a function of the number of jars of peanut butter purchased j.

3. **Identify the independent and dependent variables.**

Since the total number of peanuts depends on the number of jars of peanut butter, the number of peanuts $p(j)$ is the dependent variable and the jars of peanut butter j is the independent variable.

4. **What values of the domain and range make sense for this situation? Explain.**

Only whole numbers make sense for the domain because you cannot buy a fraction of a jar. The range values depend on the domain values, so the range will be multiples of 770.

5. **Write a function to represent the total number of peanuts. Then determine the number of peanuts in 7 jars of peanut butter.**

Words	The number of peanuts	equals	770 times	the number of jars
Function	$p(j)$	=	770 ·	j

The function $p(j) = 770j$ represents the situation.

To find the number of peanuts in 7 jars of peanut butter, substitute 7 for *j*.

$p(j) = 770j$ Write the function.

$p(j) = 770(7)$ or 5,390   Substitute 7 for *j*.

There are 5,390 peanuts in 7 jars of peanut butter.

STOP and Reflect

What are the similarities and difference among the terms domain, range, independent variable, and dependent variable? Explain below.

Got it? Do these problems to find out.

d. _____

A scrapbooking store is selling rubber stamps for $4.95 each. The total sales f(n) is a function of the number of rubber stamps n sold.

d. Identify the independent and dependent variables.

e. What values of the domain and range make sense for this situation? Explain.

e. _____

f. Write a function to represent the total sales. Then determine the total sales for 5 stamps.

f. _____

Guided Practice

1. Find $f(4)$ if $f(x) = x - 6$. (Example 1) _____

2. Choose four values for x to make a function table for $f(x) = 8 - x$.
Then state the domain and range of the function. (Example 2)

x	$8 - x$	$f(x)$

_____

3. A hot air balloon can hold 90,000 cubic feet of air. It is being inflated at a rate of 6,000 cubic feet per minute. The total cubic feet of air $a(t)$ is a function of the time in minutes t. (Examples 3–5)

a. Identify the independent and dependent variables.

_____

b. What values of the domain and range make sense for

this situation? Explain. _____

_____

c. Write a function to represent the total amount of air.
Then determine the total amount of air in 6 minutes.

_____

4. Ⓔ **Building on the Essential Question** How does the domain affect the range in a function?

_____

_____

_____

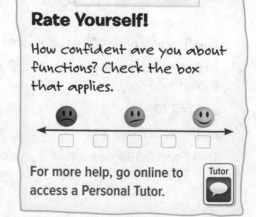

Rate Yourself!

How confident are you about functions? Check the box that applies.

For more help, go online to access a Personal Tutor.

Independent Practice

Go online for Step-by-Step Solutions

Find each function value. (Example 1)

1. $f(7)$ if $f(x) = 5x$ _____

2. $f(9)$ if $f(x) = x + 13$ _____

3 $f(4)$ if $f(x) = 3x - 1$ _____

Show your work.

Choose four values for x to make a function table for each function. Then state the domain and range of the function. (Example 2)

4. $f(x) = 6x - 4$

x	$6x - 4$	$f(x)$

5 $f(x) = 5 - 2x$

x	$5 - 2x$	$f(x)$

6. $f(x) = 7 + 3x$

x	$7 + 3x$	$f(x)$

7. In a recent 82-game season, Dwight Howard of the Orlando Magic averaged 20.7 points per game. His approximate total points scored $p(g)$ is a function of the number of games played g. (Examples 3–5)

a. Identify the independent and dependent variables.

_____

_____

b. What values of the domain and range make sense for this situation? Explain.

_____

_____

c. Write a function to represent the total points scored. Then determine the number of points scored in 9 games.

_____

8. **MP Model with Mathematics** Refer to the graphic novel frame below for Exercises a–c.

a. Write a function to represent the total cost c of printing and shipping any number of pictures p. _____

b. Make a function table on a separate piece of paper to find the total cost of printing and shipping 25, 50, 75, and 100 pictures.

c. On a separate piece of paper, graph the ordered pairs on a coordinate plane. Can you determine how many pictures Brian can ship for $25?

Copy and Solve Find each function value. Show your work on a separate piece of paper.

9. $f\left(\dfrac{5}{6}\right)$ if $f(x) = 2x + \dfrac{1}{3}$

10. $f\left(\dfrac{5}{8}\right)$ if $f(x) = 4x - \dfrac{1}{4}$

H.O.T. Problems Higher Order Thinking

11. **MP Reason Abstractly** If $f(-3) = -8$, write a function rule and find the function values for zero, a negative, and a positive value of x.

_____

12. **MP Persevere with Problems** Write the function rule for each function.

a.
x	f(x)
−3	−30
−1	−10
2	20
6	60

b.
x	f(x)
−5	−9
−1	−5
3	−1
7	3

c.
x	y
−2	−3
1	3
3	7
5	11

d.
x	y
−2	−5
1	1
3	5
5	9

13. **MP Persevere with Problems** If $f(x) = 4x - 3$ and $g(x) = 8x + 2$, find each function value.

a. $f[g(3)]$ _____

b. $g[f(5)]$ _____

c. $g\{f[g(-4)]\}$ _____

Extra Practice

Find each function value.

14. $f(-12)$ if $f(x) = 2x + 15$

Homework Help ➡

$f(x) = 2x + 15$

$f(-12) = 2(-12) + 15$

$f(-12) = -24 + 15$

$f(-12) = -9$

15. $f(-7)$ if $f(x) = 8x + 15$

_____

16. $f(9)$ if $f(x) = 5x - 16$

_____

Choose four values for x to make a function table for each function. Then state the domain and range of the function.

17. $f(x) = x - 9$

x	x − 9	f(x)

_____

_____

18. $f(x) = 7x$

x	7x	f(x)

_____

_____

19. $f(x) = 4x + 3$

x	4x + 3	f(x)

_____

_____

20. A photographer takes an average of 15 pictures per session. The total number of pictures $p(s)$ is a function of the number of sessions s.

a. Identify the independent and dependent variables.

_____

_____

b. What values of the domain and range make sense for this situation?

Explain. _____

_____

c. Write a function to represent the total number of pictures taken. Then determine the number of pictures taken in 22 sessions.

_____

21. **MP** **Reason Abstractly** Leon belongs to a music club that charges a monthly fee of $5, plus 0.50 per song that he downloads. Write a function to represent the amount of money $m(s)$ he would pay in one month to download s songs. What is the cost if he downloads 30 songs?

_____

22. A store is having a 25% off sale, and Manuel has a coupon good for $10 off his total purchase. The function $f(x) = 0.75x - 10$ represents the final cost of an item that costs x dollars after the discount and coupon are applied.

$5	$50
$20	$60
$35	$75
$40	$80

Select values to complete the function machine for the regular prices and sale prices of items A, B, C, and D.

Input A: $60

Input B: [ ]

Input C: $20

Input D: [ ]

➡ $f(x) = 0.75x - 10$ ➡

Output A: [ ]

Output B: $20

Output C: [ ]

Output D: $50

23. Stephanie received a $25 gift card to an online music store. The cost of purchasing one song is $0.95. Complete the table to show the balance remaining on Stephanie's gift card $n(s)$ after purchasing s songs.

Number of Songs s	Balance Remaining $n(s)$
2	
4	
5	
8	
10	

24. Cheryl is training for a marathon. She runs about 85 miles per week. **6.EE.7**

a. Write an equation to find the total miles m run in any number of weeks w. _____

b. Complete the table to find the total miles run in 3, 4, 5, or 6 weeks.

Evaluate each expression if $p = 5$ and $q = 12$. **6.EE.2**

25. $\dfrac{3p - 6}{8 - p}$ _____

26. $\dfrac{4q}{q + 2(p + 1)}$ _____

27. $\dfrac{q \cdot q}{4p - 2}$ _____

Lesson 4
Linear Functions

 ## Real-World Link

Up, Up, and Away The Lockheed SR-71 Blackbird has a top speed of 36.6 miles per minute. If x represents the minutes traveled at this speed, the function rule for the distance traveled is $y = 36.6x$.

1. Complete the function table.

Input	x	1	2	3	4
Rule	$36.6x$	$36.6(1)$	$36.6(2)$		
Output	y	36.6			
(Input, Output)	(x, y)	$(1, 36.6)$			

2. Graph the ordered pairs (x, y) on the coordinate plane provided. What do you notice about the graph?

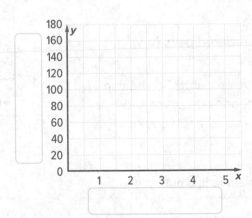

 ## Essential Question

HOW can we model relationships between quantities?

Vocab ## Vocabulary

linear function
continuous data
discrete data

 CCSS ## Common Core State Standards

Content Standards
8.F.1, 8.F.3, 8.F.4

MP Mathematical Practices
1, 3, 4, 7

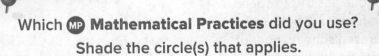

Which MP Mathematical Practices did you use?
Shade the circle(s) that applies.

① Persevere with Problems   ⑤ Use Math Tools

② Reason Abstractly   ⑥ Attend to Precision

③ Construct an Argument   ⑦ Make Use of Structure

④ Model with Mathematics   ⑧ Use Repeated Reasoning

Graph a Function

Sometimes functions are written using two variables. One variable, usually *x*, represents the domain and the other, usually *y*, represents the range. When a function is written in this form it is an equation.

Like equations, functions can be represented in words, in a table, with a graph, and as ordered pairs. The graph of a function is the set of ordered pairs consisting of an input and the corresponding output.

Example

1. The school store sells book covers for $2 each and notebooks for $1. Toni has $5 to spend. The function $y = 5 - 2x$ represents the number of book covers *x* and notebooks *y* she can buy. Graph the function. Interpret the points graphed.

Step 1 Choose values for *x* and substitute them in the function to find *y*.

x	5 − 2x	y
0	5 − 2(0)	5
1	5 − 2(1)	3
2	5 − 2(2)	1
3	5 − 2(3)	−1

Step 2 Graph the ordered pairs (*x*, *y*).

She cannot buy negative amounts. So she can buy 0 covers and 5 notebooks, 1 cover and 3 notebooks, or 2 covers and 1 notebook.

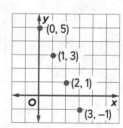

Got it? Do this problem to find out.

a. _____

a. The farmer's market sells apples for $2 per pound and oranges for $1 per pound. Marjorie has $10 to spend. The function $y = 10 - 2x$ represents the number of apples *x* and oranges *y* Marjorie can purchase. Graph the function and interpret the points graphed.

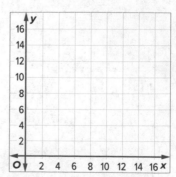

> **Function Notation**
>
> The equation $y = 5 - 3x$ can also be written in function notation as $f(x) = 5 - 3x$.

Example

2. Graph $y = x + 2$.

Step 1 Make a function table. Select any four values for the domain x. Substitute these values for x to find the value of y, and write the corresponding ordered pairs.

x	$x + 2$	y	(x, y)
0	$0 + 2$	2	$(0, 2)$
1	$1 + 2$	3	$(1, 3)$
2	$2 + 2$	4	$(2, 4)$
3	$3 + 2$	5	$(3, 5)$

Step 2 Graph each ordered pair. Draw a line that passes through each point.

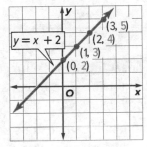

The line is the complete graph of the function. The ordered pair corresponding to any point on the line is a solution of the equation $y = x + 2$.

Solutions
The solutions of an equation are ordered pairs that make an equation representing the function true.

Check It appears that $(-2, 0)$ is also a solution. Check this by substitution.

$y = x + 2$ Write the function.

$0 \stackrel{?}{=} -2 + 2$ Replace x with -2 and y with 0.

$0 = 0 \checkmark$ Simplify.

Got it? Do these problems to find out.

b. $y = x - 5$

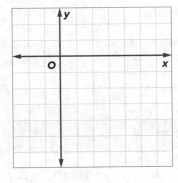

c. $y = -2x$

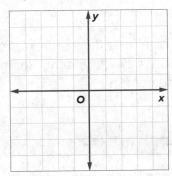

Representing Functions

Words   The value of *y* is one less than the corresponding value of *x*.

Equation $y = x - 1$ **Ordered Pairs**   (0, −1), (1, 0), (2, 1), (3, 2)

Table

x	y
0	−1
1	0
2	1
3	2

Graph

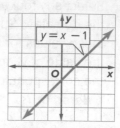

Continuous and Discrete

If the domain of a function is integers, this is an example of a discrete function. If the domain is all real numbers, this is an example of a continuous function.

A **linear function** is a function in which the graph of the solutions forms a straight line. Therefore, an equation of the form $y = mx + b$ is a *linear function*.

A function can be considered continuous or discrete. **Continuous data** can take on any value, so there is no space between data values for a given domain. **Discrete data** have space between possible data values. Graphs of continuous data are represented by solid lines and graphs of discrete data are represented by dots.

Continuous Data	Discrete Data
the number of ounces in a glass	the number of glasses in a cupboard
the weight of each chocolate chip	the number of chocolate chips in a bag

You can determine if data that model real-world situations are discrete or continuous by considering whether all numbers are reasonable as part of the domain.

Examples

Each person that enters a store receives a coupon for $5 off his or her entire purchase.

3. **Write a function to represent the total value of the coupons given out.**

Let *y* represent the total value of the coupons and *x* represent the number of people. The function is $y = 5x$.

4. Make a function table to find the total value of the coupons given out to 5, 10, 15, and 20 customers.

x	5x	y
5	5(5)	25
10	5(10)	50
15	5(15)	75
20	5(20)	100

STOP and Reflect

Explain below how a function table can be used to graph a function.

5. Graph the function. Is the function continuous or discrete? Explain.

Use the ordered pairs from the function table to graph the function.

There can only be a whole number amount of customers. The function is discrete. So, the points are not connected.

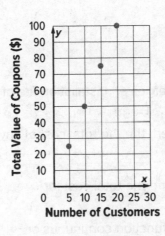

Got it? Do these problems to find out.

Show your work.

A store sells assorted nuts for $5.95 per pound.

d. Write a function to represent the total cost of any number of pounds of nuts.

e. Complete the function table below to find the total cost of 1, 2, 3, 4, or 5 pounds of nuts.

f. Graph the function. Is the function continuous or discrete? Explain.

d. _____

f. _____

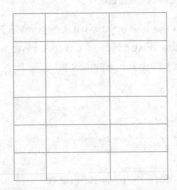

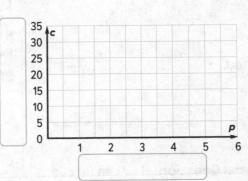

Graph each function. (Example 2)

1. $y = x + 5$

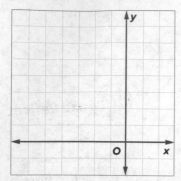

2. $y = 3x - 2$

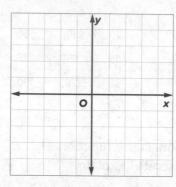

3. A satellite cable company charges an installation fee of $50 plus an additional $35.95 per month for service. (Examples 1, 3–5)

a. Write a function to represent the the total cost of any number of months of service. _____

b. Make a function table to find the total cost for 1, 2, 3, 4, or 5 months.

c. Graph the function. Is the function continuous or discrete? Explain.

_____

_____

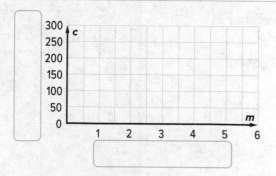

d. Interpret the points graphed. _____

_____

_____

4. @ **Building on the Essential Question** How can functions be used to solve real-world situations?

_____

_____

FOLDABLES Time to update your Foldable!

Independent Practice

eHelp
Go online for Step-by-Step Solutions

Graph each function. (Example 2)

1. $y = 4x$

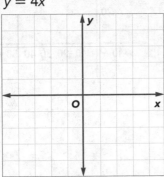

2. $y = -3x$

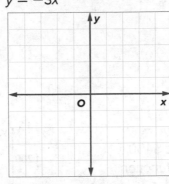

3. $y = x - 3$

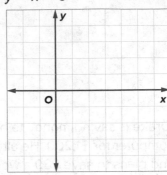

4. Financial Literacy Manuel is saving money for college. He already has $250. He plans to save another $50 per month. (Examples 1, 3–5)

a. Write a function to represent his savings for any number of months. _____

b. Make a function table to find his total savings for 2, 4, 6, 8 and 10 months.

c. Graph the function. Is the function continuous or discrete? Explain. _____

_____

d. Interpret the points graphed. _____

_____

_____

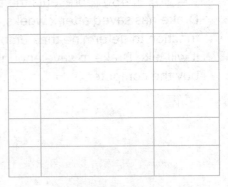

Manuel's Savings

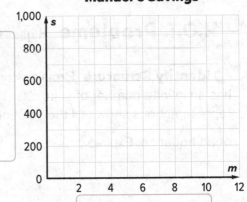

5. Copy and Solve The table shows the cost to rent different items.

a. Write a function to represent each situation.

Item	Deposit ($)	Cost per Hour ($)
Mountain bike	15	4.25
Scooter	25	2.50

_____

b. On a separate piece of paper, make a function table to find the total cost to rent each item for 2, 3, 4, or 5 hours.

c. On a separate piece of grid paper, graph the functions on the same coordinate plane. Are the functions continuous or discrete? Explain.

_____

d. Will the mountain bike or the scooter cost more to rent for 8 hours?

_____

e. How much is the cost to rent the mountain bike for 8 hours? _____

6. **MP Model with Mathematics** The formula $F = 1.8C + 32$ compares temperatures in degrees Celsius C to temperatures in degrees Fahrenheit F. Find four ordered pairs (C, F) that are solutions of the equation. Then graph the function.

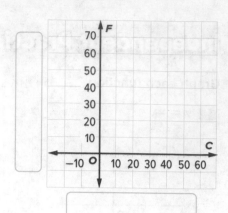

7. Drake is saving money to buy a new computer for $1,200. He already has $450 and plans to save $30 a week. The function $y = 30x + 450$ represents the amount Drake has saved after x weeks. Graph the function to determine the number of weeks it will take Drake to save enough money to buy the computer.

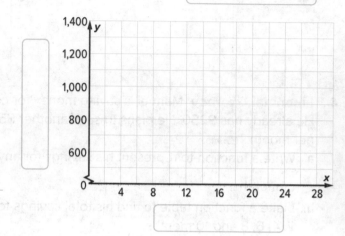

H.O.T. Problems Higher Order Thinking

8. **MP Identify Structure** Explain why a linear function that is continuous has an infinite number of solutions. Then determine which of the following representations shows all the solutions of the function: a table, a graph, or an equation. Explain. _____

_____

_____

9. **MP Persevere with Problems** Name the coordinates of four points that satisfy the linear function shown. Then give the function rule.

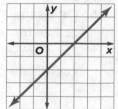

10. **MP Model with Mathematics** Write a set of four ordered pairs that represents a linear function. Then give the function rule.

_____

Extra Practice

11. A store sells T-shirts x in packs of 5 and regular shirts y individually. Graph the function $y = 10 - 5x$ to determine the number of each type of shirt Bethany can have if she buys 10 shirts.

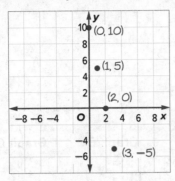

She cannot buy negative amounts. So, she can buy 0 T-shirt packs and 10 shirts individually, 1 T-shirt pack and 5 shirts individually, or 2 T-shirt packs and 0 shirts individually.

12. Fancy goldfish x cost $3 each and common goldfish y cost $1 each. Graph the function $y = 20 - 3x$ to determine how many of each type of goldfish Tasha can buy for $20.

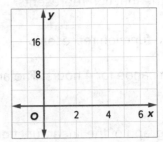

_____

_____

_____

_____

_____

Graph each function.

13. $y = 3x - 7$

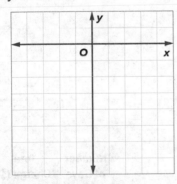

14. $y = 2x + 3$

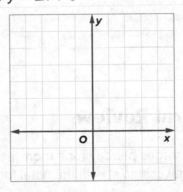

15. $y = \frac{1}{3}x + 1$

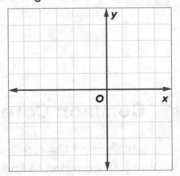

16. **MP Model with Mathematics** The equation $y = 1.09x$ describes the approximate number of yards y in x meters.

a. Would negative values of x have any meaning in this situation? Explain.

_____

b. Graph the function.

c. About how many meters is a 40-yard race?

_____

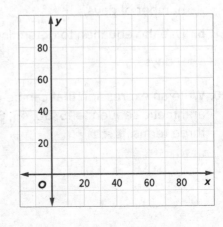

17. A rental company charges $8 plus $5.50 per hour to rent a canoe. Fill in each box below to make true statements.

   a. A function that can be used to find the total cost y of renting a canoe for x hours is [   ].

   b. The domain of the function represents [   ].

   c. The range of the function represents [   ].

   d. Michelle rents a canoe for 4 hours. The amount she pays is [   ].

18. Select the correct graph of the linear function shown in each table below.

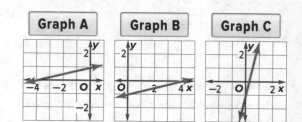

Graph A   Graph B   Graph C

Function				Graph
x	−4	0	4	
y	−2	−1	0	
x	−1	0	1	
y	−5	−1	3	
x	−4	0	4	
y	0	1	2	

Common Core Spiral Review

19. The table shows the rate a hotel charges per day for a room. **6.EE.7**

   a. Write an equation to find the total cost to stay in the hotel for any number of days. _____

   b. Use the equation to determine how much it would cost to stay for 9 days. _____

Number of Days	Total Charge ($)
1	65
2	130
3	195
4	260

20. Write an expression that can be used to find the nth term of the arithmetic sequence 15, 30, 45, 60, ... Then write the next three terms. **6.EE.2**

_____

Make a Table

CCSS Content Standards
8.F.4

MP Mathematical Practices
1, 2, 4

Case #1 Play Catch Up

Emilio's family is going on vacation. His mom and sister leave at 7:00 in the morning, driving an average of 45 miles per hour. Emilio and his dad leave at 8:00. His dad drives an average of 60 miles per hour.

Will Emilio and his dad catch up to his mom and sister?

Understand *What are the facts?*

You know the times they left and their rates. You need to know if Emilio and his dad will catch up to his mom and sister.

Plan *What is your strategy to solve this problem?*

Make a table that shows how many miles each driver has driven.

Solve *How can you apply the strategy?*

Hours Since 7:00 A.M.	Distance Traveled (mi)	
	Emilio's Mom	Emilio's Dad
0	0	0
1	45	0
2		60
3		
4		

At [ ] A.M., Emilio and Emilio's dad will catch up to his mom and sister.

Check *Does the answer make sense?*

45 mph × [ ] h = [ ] mi   60 mph × [ ] h = [ ] mi

The distances are equal. ✔

Analyze the Strategy [Tutor]

MP Reason Abstractly Suppose Emilio's mom drives at an average speed of 50 miles per hour. At what time will Emilio and his dad catch up to her?

Rina wants to rent a karaoke machine for a family reunion. The prices to rent the machine from two different companies are shown. For how many days must she rent the machine for the cost from each place to be the same?

Company	Deposit	Cost Per Day
Mike's Music	$5	$1.25
Karaoke Korner	$4	$1.50

Understand

Read the problem. What are you being asked to find?

I need to find _____.

Underline key words and values. What information do you know?

I know that Mike's Music has a deposit of ⬜ and charges ⬜ a day.

Karaoke Korner's deposit is ⬜ and they charge ⬜ a day.

Plan

Choose a problem-solving strategy.

I will use the _____ strategy.

Solve

Use your problem-solving strategy to solve the problem.

	Day 1	Day 2	Day 3	Day 4
Mike's Music				
Karaoke Korner				

So, the cost at both companies is the same at _____.

Check

Use information from the problem to check your answer.

Mike's Music charges ⬜ or ⬜ for the first day.

Each day adds another ⬜.

Karaoke Korner charges ⬜ or ⬜ for the first day.

Each day adds another ⬜.

At ⬜ days, both companies charge ⬜.

Work with a small group to solve the following cases.
Show your work on a separate piece of paper.

Case #3 Plants

The table shows the height of a giant bamboo plant for Days 5-9.
The bamboo grew at a steady rate each day, starting on Day 5.

From Day 5 to Day 9, the bamboo plant grew about what percent of its final height?

Bamboo Growth	
Number of Days	Total Growth (ft)
5	?
6	?
7	10
8	13.5
9	17

Case #4 Financial Literacy

Sophia and Scott each open a bank account with an initial deposit of $50 each. Scott planned to save 30% of his earnings from his after-school job. The job pays $8 per hour, and he works 25 hours each week. For four weeks, Sophia saved $45 per week. After that, she saved an additional $30 per week.

During which week will they have the same amount of money in their bank accounts?

Case #5 Fitness

Marcie created her own fitness plan to increase the amount of time she spends exercising each week. The total number of minutes in her exercise plan for each week is shown.

If she continues this pattern, for how many hours will she exercise in the 8th week?

Week	Exercise (min)
1	35
2	50
3	80
4	125
5	185

Use any strategy!

Case #6 Animals

The graph shows the maximum length of several animals. The maximum length of a walrus is twice the maximum length of a lion, which is 0.4 meter longer than the maximum length of a giant panda.

Find the maximum length of a walrus.

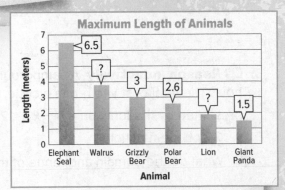

Maximum Length of Animals

Length (meters) vs Animal

Animal	Length
Elephant Seal	6.5
Walrus	?
Grizzly Bear	3
Polar Bear	2.6
Lion	?
Giant Panda	1.5

Mid-Chapter Check

Vocabulary Check

1. **MP** **Be Precise** Define *linear equation*. Give an example of a linear equation. (Lesson 1)

   _____

   _____

2. Describe the difference between the graph of a set of discrete data and the graph of a set of continuous data. (Lesson 4) _____

   _____

Skills Check and Problem Solving

3. There are 20 nickels in one dollar. (Lesson 2)

   a. Write an equation to find the number of nickels *n* in any

    number of dollars *d*. _____

   b. Make a table to find the number of nickels in 5, 10, 15, or 20 dollars: Then graph the ordered pairs.

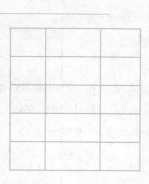

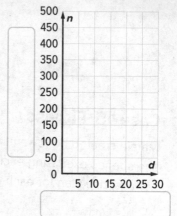

Find each function value. (Lesson 3)

4. $f(8)$ if $f(x) = 15x$

   _____

5. $f(2)$ if $f(x) = 2x - 5$

   _____

6. $f(4)$ if $f(x) = -3x + 15$

   _____

7. **MP** **Reason Inductively** A campground rents bicycles by the hour. The total cost *y* to rent a bicycle, including deposit, is presented by the function $y = \frac{1}{3}x + 12$. (Lessons 2 and 4)

   a. Graph the function.

   b. What do the domain and range of the function represent?

   _____

   _____

   c. Is the function continuous or discrete? _____

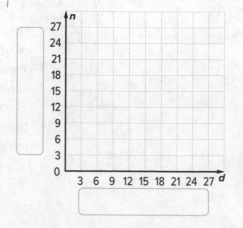

Compare Properties of Functions

 Real-World Link

Science Museum Carlos' annual membership to the science museum can be represented by the function $c = 29.99$, where c represents the cost in dollars. The cost for Stephanie to pay per visit is shown in the table.

Visits	Cost ($)
1	5
2	10
3	15
4	20
5	25

 Essential Question

HOW can we model relationships between quantities?

 Common Core State Standards

Content Standards 8.F.2, 8.F.4

MP Mathematical Practices 1, 2, 3, 4

1. Make a table to represent Carlos' membership.

Months	Cost ($)

2. Describe the rate of change for each function.

_____

_____

3. Who pays more for two visits? Explain.

_____

_____

4. Who pays more for six visits? Explain.

_____

_____

Which MP Mathematical Practices did you use?
Shade the circle(s) that applies.

① Persevere with Problems   ⑤ Use Math Tools

② Reason Abstractly   ⑥ Attend to Precision

③ Construct an Argument   ⑦ Make Use of Structure

④ Model with Mathematics   ⑧ Use Repeated Reasoning

Compare Two Functions

Functions can be represented by a table, graph, equation, or words. You can compare two functions represented in different forms.

Example

1. A zebra's main predator is a lion. Lions can run at a speed of 53 feet per second over short distances. The graph at the right shows the speed of a zebra. Compare their speeds.

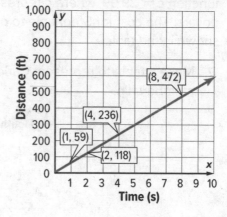

To compare their speeds, compare the rates of change.

A lion can travel at a rate of 53 feet per second.

To find the rate of change for a zebra, choose two points on the line and find the rate of change between them.

$$\frac{\text{Change in distance}}{\text{Change in time}} = \frac{118 - 59}{2 - 1} \text{ or } \frac{59}{1}$$

A zebra can travel at a rate of 59 feet per second. Since 59 > 53, the speed of a zebra is greater than the speed of a lion.

Show your work.

Got it? Do this problem to find out.

a. _____

a. A certain car has a gas mileage of 22 miles per gallon. The gas mileage of a certain sport utility vehicle is represented by the function shown. Compare their gas mileage.

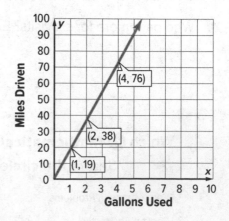

Example

Tutor

2. The function $m = 140h$, where m is the miles traveled in h hours, represents the distance traveled of the first Japanese high speed train. The distance traveled of a high speed train operating today in China is shown in the table. Assume the relationship between the two quantities is linear.

Train Rate in China

Hours	Miles
1	217
2	434
3	651

a. Compare the functions' *y*-intercepts and rates of change.

Compare the *y*-intercepts.

At 0 hours, no distance has been covered. So, the *y*-intercepts are the same, 0.

Compare the rates of change.

The speed of the Japanese train is 140 miles per hour.

Use the table to find the speed of the Chinese train.

The speed of the Chinese train is $\dfrac{217 \text{ miles}}{1 \text{ hour}}$ or 217 miles per hour.

Train Rate in China

	Hours	Miles	
+1	1	217	+217
+1	2	434	+217
	3	651	

Since 217 > 140, the function representing the Chinese high speed train has a greater rate of change than the function representing the Japanese high speed train.

b. If you ride each train for 5 hours, how far will you travel on each?

Find the distance on the Japanese train.

$m = 140h$ Write the function.

$m = 140(5)$ Replace *h* with 5.

$m = 700$ Simplify.

You will travel 700 miles in 5 hours on the Japanese train.

Find the distance on the Chinese train by extending the table.

You will travel 1,085 miles in 5 hours on the Chinese train.

Train Rate in China

	Hours	Miles	
+1	1	217	+217
+1	2	434	+217
+1	3	651	+217
+1	4	868	+217
	5	1,085	

Got it? Do these problems to find out.

The number of new movies a store receives can be represented by the function $m = 7w + 2$, where m represents the number of movies and w represents the number of weeks. The number of games the same store receives is shown in the table.

Week	Number of New Games
1	3
2	6
3	9

b. Compare the functions' *y*-intercepts and rates of change.

c. How many new movies and games will the store have in Week 6?

Example

Tutor

3. **Financial Literacy** Angela and Benjamin each have a monthly cell phone bill. Angela's monthly cell phone bill is represented by the function $y = 0.15x + 49$, where *x* represents the minutes and *y* represents the cost. Benjamin's monthly cost is shown in the graph.

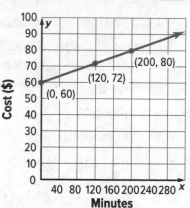

a. **Compare the *y*-intercepts and rates of change.**

The function for Angela's bill has a *y*-intercept of 49. You can see from the graph that the function for Benjamin's bill has a *y*-intercept of 60. So, Benjamin has a greater initial cost.

The rate of change for Angela's monthly bill is $0.15 per minute. Find the rate of change for Benjamin's bill.

$$\frac{\text{change in cost}}{\text{change in minutes}} = \frac{80 - 60}{200 - 0} \text{ or } 0.10$$

The rate of change for Benjamin's bill is $0.10 per minute. So, Angela pays more per minute than Benjamin.

b. **What will be the monthly cost for Angela and Benjamin for 200 minutes?**

Angela's monthly cost is represented by $y = 0.15x + 49$. At 200 minutes, Angela will pay $0.15(200) + 49$ or $79.

Use the graph to find Benjamin's cost. At 200 minutes, Benjamin will pay $80.

Got it? Do these problems to find out.

d. _____

Financial Literacy Mandy and Sarah each have a membership to the gym. Mandy's membership is represented by the function $y = 3x + 29$, where x represents the hours with a trainer and y represents the cost. The cost of Sarah's membership is shown in the graph.

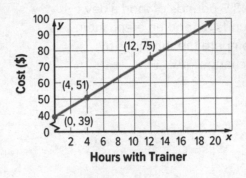

d. Compare the y-intercepts and rates of change.

e. What will be the total cost for Mandy and Sarah if they each have 4 hours with a trainer?

Show your work.

Example

e. _____

4. **Financial Literacy** Lorena's mother needs to rent a truck to move some furniture. The cost to rent a truck from two different companies is shown in the table and graph. Which company should she use to rent the truck for 40 miles?

Cross Town Movers

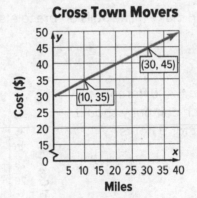

Find the cost of renting a truck from Ron's Rentals by extending the table.

After 40 miles, the cost will be $75 + $25 or $100.

Ron's Rentals	
Miles	**Cost ($)**
10	25
20	50
30	75

+10 ( ) +25

> **Multiple Representations**
> You can find the two costs of truck rentals by extending the table, extending the line on the graph, or writing an equation. The method you use will depend on the information that you are given.

Find the cost of renting a truck from Cross Town Movers by analyzing the graph. The y-intercept of the graph is 30.

The slope or rate of change is $\frac{45 - 35}{30 - 10}$ or 0.5. The equation $y = 0.5x + 30$ where y represents the total cost and x represents the miles driven can be used to find the total cost of renting the truck. After 40 miles, the cost will be $0.5(40) + 30$ or $50.

So, Cross Town Movers would cost less for 40 miles.

1. A tiger in captivity is fed 13.5 pounds of food a day. The graph shows the pounds of food an elephant in captivity eats per day. Compare the functions by comparing their rates of change. (Example 1)

 Show your work.

_____

_____

_____

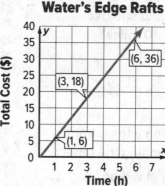

Pounds of Food vs Time (days)

Points: (1, 125), (2, 250), (3, 375), (4, 500), (5, 625)

2. Carol's profit at a craft fair is represented by the function $p = 5b - 15$, where p is the profit and b is the number of bracelets she sells. Kate's profit is shown in the table. (Examples 2 and 3)

   a. Compare the y-intercepts and rates of change.

   _____

   _____

   b. How much will each girl make if she sells 30 bracelets?

   _____

Bracelets Sold	Profit ($)
1	5
2	10
3	15
4	20

3. The cost to rent a raft from two different companies is shown. Which company should you use if you rent the raft for 9 hours?

   (Example 4) _____

   _____

   _____

Water's Edge Rafts

Total Cost ($) vs Time (h)

Points: (1, 6), (3, 18), (6, 36)

Ryan's Rafts	
Time (h)	Total Cost ($)
1	15.00
2	17.25
3	19.50
4	21.75
5	24.00

4. **Building on the Essential Question** What are the advantages and disadvantages to representing a function as an equation instead of a graph? _____

_____

_____

_____

_____

Rate Yourself!

Are you ready to move on?
Shade the section that applies.

YES   ?   NO

For more help, go online to access a Personal Tutor.

Independent Practice

eHelp
Go online for Step-by-Step Solutions

1 For the first leg of the Ramirez family's trip, their speed averages 68 miles per hour. The second leg is shown in the graph. Compare the speeds for each part of their trip. (Example 1)

 Show your work.

_____

_____

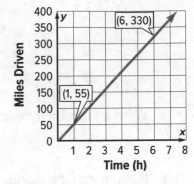

2. The late fees for a school library are represented by the function $c = 0.25d$, where c is the total cost and d is the number of days a book is late. The fees charged by a city library are shown in the table. (Examples 2 and 3)

Days Late	1	2	3
Cost ($)	0.35	0.70	1.05

  a. Compare the functions' y-intercepts and rates of change.

  _____

  _____

  b. Shamar checks out one book at each library and returns both books

  3 days late. What are the late fees for each library? _____

  _____

3 Matt and Seth purchase baseball cards each week. The amount of cards they each have in their collection is shown in the graph and table. Who will have more cards in Week 20? Justify your response.

(Example 4)

Seth's Collection

Week	Number of Cards
1	4
2	8
3	12

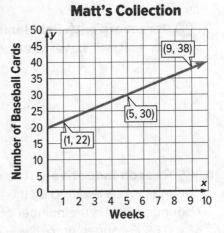

Matt's Collection

_____

_____

4. The Shaw family is building a patio. One person can place the flagstone at a rate of 4.5 per hour. The equation $s = 11h$ represents the number of stones s that two people can place in h hours. How many more flagstones can 2 people place in 3 hours than one person? Explain.

_____

_____

5. **MP Reason Abstractly** Refer to the conversions in the tables below.

Cups	Ounces
1	8
2	16
3	24
4	32

Pints	Ounces
1	16
2	32
3	48
4	64

Quarts	Ounces
1	32
2	64
3	96
4	128

a. Write a function for each table.

_____

b. If you graph the points, the graph for which function would have the steepest slope? Justify your response.

_____

_____

c. Which function has the least rate of change? Explain.

_____

H.O.T. Problems Higher Order Thinking

6. **MP Model with Mathematics** Write a real-world problem where you would want to compare rates of change for two different functions.

_____

7. **MP Persevere with Problems** Explain why the graph of the function $y = 3x + 40$ will never intersect the graph of the function $y = 3x + 35$.

_____

_____

8. **MP Reason Inductively** The exchange rate to convert U.S. dollars to British pounds is represented by the function $p = 0.61d$ where p is the amount in pounds and d is the number of dollars. One U.S. dollar can also be exchanged for 0.69 European euro. If you exchange $250 for pounds and $250 for euros, which of the following is true? _____

I   You will receive 152.5 pounds and 172.5 euros.

II You will receive about 410 pounds and about 362 euros.

III You will receive 250 dollars.

IV You will receive the same amount of pounds and euros.

Extra Practice

9. A fabric store sells cotton for $7.00 a yard. The price of special occasion fabric is shown in the graph. Compare the functions' rates of change.

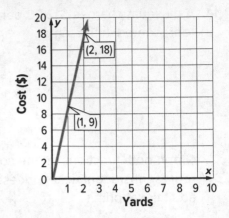

Cotton fabric: $7.00 per yard

Special occasion fabric: $\frac{18-9}{2-1} = \frac{9}{1}$ or $9.00 per yard.

The special occasion fabric has the greater rate of change.

10. Two players played a game. The first player's score is represented by the function $p = 5c - 3$, where p is the number of points scored and c is the number of correct answers. The second player's score is shown in the table.

a. Compare the functions by comparing their y-intercepts and rates of change. _____

_____

_____

Questions Answered	Score
1	5
2	10
3	15
4	20

b. How many points will the first player have if he or she correctly answers 30 questions? _____

11. **MP** **Justify Conclusions** Jesse and Juan each open savings accounts. The amounts in Jesse's account are shown in the table. Juan saves $5 per week. Who will have more saved in 8 weeks? Explain.

_____

_____

12. Canada Olympic Park features sports training and entertainment facilities. The Monster zip line produces average speeds of 120 kilometers per hour. A smaller line produces speeds represented by the function $d = 50h$ where d is the distance in kilometers after h hours. How much farther could you travel on the Monster zip line in 0.25 hours?

Jesse's Savings	
Week	Amount Saved ($)
1	16
2	19
3	22
4	25
5	28

_____

13. Raj gets a 1.5 mile head start and runs at a rate of 4.5 miles per hour. Jacinda's progress is represented by a graph that goes through the points (1, 10), (2, 20), and (3, 30). How long will Jacinda need to run to catch up with Raj? _____

14. A museum charges $12.50 per adult ticket. The price of a student ticket is represented in the table.

Student Ticket Price			
Tickets	1	2	3
Price ($)	8.50	17.00	25.50

Graph both functions on the coordinate plane. Use circles to represent student ticket prices and use squares to represent adult ticket prices.

Compare the rates of change and *y*-intercepts of the linear functions. Explain what each of these represent.

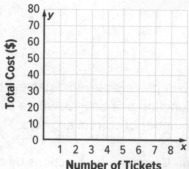

15. Julie swam, biked, and ran in a 22.2 mile triathlon. She completed the race in 2.15 hours. The function $m = 13.8h$ represents the miles *m* Julie biked in *h* hours. Was her average speed biking less than or greater than her average speed for the entire race? Justify your answer. Round to the nearest tenth if necessary.

Write an equation in slope-intercept form for each table of values. 8.F.3

16.

x	−1	0	1	2
y	−7	−3	1	5

_____

17.

x	−3	−1	1	3
y	7	5	3	1

_____

Construct Functions

Real-World Link

Parties Dylan is planning to have his birthday party at a skating rink. The rink charges a party fee plus an additional charge for each guest.

Number of Guests, x	Total Cost ($), y
1	53
2	56
3	59
4	62
5	65
6	68

1. Choose two points from the table and find the rate of change.

_____

_____

2. Write a function to represent this situation.

_____

_____

3. Graph the ordered pairs. Then extend the line of the graph until it crosses the y-axis.

4. Use the function to find the amount the skating rink charges for the party fee.

_____

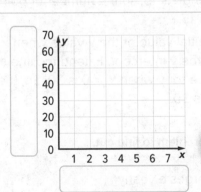

whoa!

 Essential Question

HOW can we model relationships between quantities?

 Common Core State Standards

Content Standards
8.F.4

MP Mathematical Practices
1, 3, 4

Which MP Mathematical Practices did you use?
Shade the circle(s) that applies.

① Persevere with Problems

⑤ Use Math Tools

② Reason Abstractly

⑥ Attend to Precision

③ Construct an Argument

⑦ Make Use of Structure

④ Model with Mathematics

⑧ Use Repeated Reasoning

Analyze Graphs, Words, and Tables

The *initial value of a function* is the corresponding *y*-value when *x* equals 0. You can find the initial value of a function from graphs, words, and tables.

Example

Tutor

1. A shoe store offers free points when you sign up for their rewards card. Then, for each pair of shoes purchased, you earn an additional number of points. The graph shows the total points earned for several pairs of shoes. Find and interpret the rate of change and initial value.

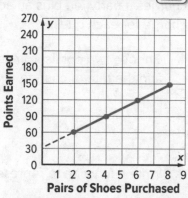

To find the rate of change, choose two points from the graph.

$$\frac{\text{change in points}}{\text{change in pairs}} = \frac{(90-60)\text{ points}}{(4-2)\text{ pairs}}$$

$$= \frac{15\text{ points}}{1\text{ pair}}$$

The rate of change is 15, so the number of points earned per pair of shoes is 15.

Next find the initial value or the *y*-value when *x* = 0. Recall this value is called the *y*-intercept. Extend the line so it intersects the *y*-axis. The value for *y* when *x* = 0 is 30. So, the initial number of points earned is 30.

Show your work.

Got it? Do this problem to find out.

a. _____

a. Music Inc. charges a yearly subscription fee plus a monthly fee. The total cost for different numbers of months, including the yearly fee, is shown in the graph. Find and interpret the rate of change and initial value.

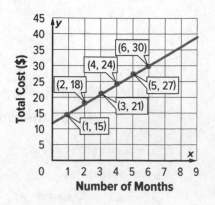

Example

Tutor

2. Joan has some photos in her photo album. Each week she plans to add 12 photos. Joan had 120 photos after 8 weeks. Assume the relationship is linear. Find and interpret the rate of change and initial value.

Since each week Joan adds 12 photos to her photo album the rate of change is 12. To find the initial value, use slope-intercept form to find the *y*-intercept.

$$y = mx + b$$ Slope-intercept form

$$y = 12x + b$$ Replace *m* with the rate of change, 12.

$$120 = 12(8) + b$$ Replace *y* with 120 and *x* with 8

$$24 = b$$ Solve for *b*.

The *y*-intercept is 24. So, the initial number of photos is 24.

Got it? Do this problem to find out.

b. A zoo charges a rental fee plus $2 per hour for strollers. The total cost of 5 hours is $13. Assume the relationship is linear. Find and interpret the rate of change and initial value.

Show your work.

b. _____

Example

Tutor

3. The table shows how much money Ava has saved. Assume the relationship between the two quantities is linear. Find and interpret the rate of change and initial value.

Number of Months, *x*	Money Saved ($), *y*
3	110
4	130
5	150
6	170

Choose any two points from the table to find the rate of change. The rate of change is $\frac{150 - 110}{5 - 3}$ or 20, so Ava saves $20 each month. To find the initial value, use the slope-intercept form to find the *y*-intercept.

$$y = mx + b$$ Slope-intercept form

$$y = 20x + b$$ Replace *m* with the rate of change, 20.

$$110 = 20(3) + b$$ Use the point (3, 110). *x* = 3, *y* = 110

$$50 = b$$ Solve for *b*.

The *y*-intercept is 50, so Ava had initially saved $50.

c. _____

Got it? Do this problem to find out.

c. The table shows the monthly cost of sending text messages. Assume the relationship between the two quantities is linear. Find and interpret the rate of change and intial value.

Number of Messages, x	Cost ($), y
5	10.50
6	10.60
7	10.70

Guided Practice

1. As part of a grand opening, an arcade gave out free tokens to the first 100 customers. The graph shows the number of tokens customers received for each dollar spent at the Play More Arcade. Find and interpret rate of change and the initial value. (Example 1)

 Show your work.

_____

_____

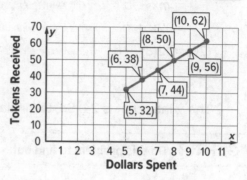

2. A historic museum charges a rental fee plus $2 per hour for an audio tour guide. The total cost for 4 hours is $12. Find and interpret the rate of change and initial value. (Example 2)

_____

_____

3. A science center charges an initial membership fee. The total cost of the membership depends on the number of people on the membership as shown in the table. Assume the relationship between the two quantities is linear. Find and interpret the rate of change and the initial value. (Example 3) _____

Number of People, x	2	3	4	5
Additional Cost ($), y	65	80	95	110

_____

Rate Yourself!

☐ I understand how to construct functions.

▶▶ Great! You're ready to move on!

4. **ⓔ Building on the Essential Question** How is the initial value of a function represented in a table and in a graph?

_____

_____

_____

☐ I still have questions about constructing functions.

📖 No Problem! Go online to access a Personal Tutor. Tutor

Independent Practice

Go online for Step-by-Step Solutions eHelp

1 A teacher read part of a book to a class. The graph shows the number of pages read by the teacher over the next several days. Find and interpret the rate of change and the initial value. (Example 1)

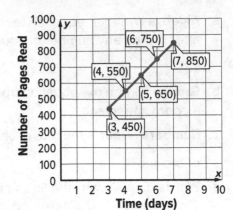

_____

_____

Show your work.

2. A water park charges a rental fee plus $1.50 per hour to rent inflatable rafts. The total cost to rent a raft for 6 hours is $15. Assume the relationship is linear. Find and interpret the rate of change and the Initial

value. (Example 2) _____

3 A teacher already had a certain number of canned goods for the food drive. Each day of the food drive, the class plans to bring in 10 cans. The total number of canned goods for day 10 is 205. Assume the relationship is linear. Find and interpret the rate of change and the initial value. (Example 2)

4. Melissa frosted some cupcakes in the morning for a party. The table shows the total number of cupcakes frosted after she starts up after lunch. Assume the relationship between the two quantities is linear. Find and interpret the rate of change and the initial value. (Example 3)

Time (min), x	5	10	15	20
Number of Cupcakes, y	28	32	36	40

_____

5. Jonas has a certain number of DVDs in his collection. The table shows the total number of DVDs in his collection over several months. Assume the relationship between the two quantities is linear. Find and interpret the rate of change and the initial value. (Example 3)

Month, x	3	6	9	12
Number of DVDs, y	18	27	36	45

6. **MP Multiple Representations** The Coughlin family is driving from Boston to Chicago. The total distance of the trip is 986 miles and each hour they will drive 65 miles.

   a. **Algebra** Write an equation to represent the number of remaining miles y after driving any number of hours x.

   _____

   b. **Graphs** Graph the equation from part **a** on a coordinate plane.

   c. **Numbers** What is the rate of change and y-intercept

   of the line? _____

   d. **Words** Explain why the line *slopes down* by 65 for

   each hour. _____

   e. **Words** Why does the line cross the y-axis at 986?

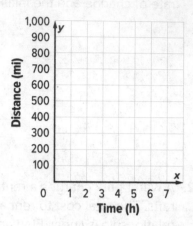

H.O.T. Problems Higher Order Thinking

7. **MP Persevere with Problems** Explain why a horizontal line has a rate of

   change of zero. _____

   _____

   _____

8. **MP Model with Mathematics** Write and solve a real-world problem in which you need to find the initial value of a function. Then explain to a classmate how you solved your problem.

   _____

   _____

9. **MP Justify Conclusions** Your teacher asks you to write a linear equation for a function in which the rate of change is -7 and the initial value is -2. You wrote the equation $y = -7x + (-2)$. Your classmate wrote the equation $y = -7x - 2$. Another classmate wrote the equation $y = (-2) + (-7)x$. Your teacher wrote the equation $y = -2 - 7x$. Who is correct? Justify your response.

   _____

   _____

   _____

Extra Practice

10. A snowboard instructor charges an initial fee plus $40 per hour for private snowboarding lessons. Carmen paid $265 for six hours of instruction. Assume the relationship is linear. Find and interpret the rate of change and initial value.

The instructor charges $40 per hour. The initial fee is $25.

Since the instructor charges $40 per hour, the rate of change is 40. To find the initial value, use slope-intercept form to find the y-intercept.

 Homework Help →

$$y = mx + b$$

$$y = 40x + b$$

$$265 = 40(6) + b$$

$$25 = b$$

The y-intercept is 25. So, the initial fee is $25.

11. A family drove to their grandmother's house. After that they averaged 200 miles per day for 8 days. They drove a total of 1,880 miles over the eight days. Assume the relationship is linear. Find and interpret the rate of change and initial value.

_____

12. **MP** **Multiple Representations** Monique and Tasha are traveling on the same highway to a family reunion at a park. Monique starts out 225 miles from the park and drives 70 miles per hour. Tasha starts out 200 miles from the park and drives 65 miles per hour.

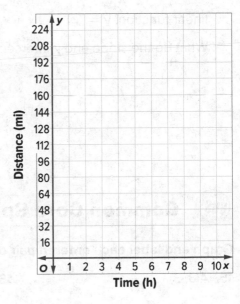

a. **Algebra** Write an equation for Monique's trip where *y* is the total distance from the park after *x* hours.

_____

b. **Algebra** Write an equation for Tasha's trip where *y* is the

total distance from the park after *x* hours. _____

c. **Graphs** Graph both equations on the same coordinate plane.

d. **Words** Will Monique's and Tasha's trip overlap before they reach the park? Explain your reasoning.

_____

_____

e. **Numbers** Interpret the initial value of each function.

_____

_____

13. The graph shows the number of gallons of water in a bathtub after filling it for a certain number of minutes. Fill in the boxes to make a true statement.

The rate of change of the function is [　　　　] per minute.

The initial value is [　　　　].

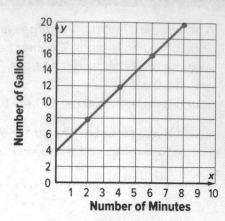

14. Coach Keller is having the last names of his players printed on their shirts. The lettering company charges a flat fee and a charge per letter. The table shows the total costs for 15, 20, 25, and 30 letters. Select the correct values to model the situation with a linear function.

Number of Letters, x	15	20	25	30
Cost ($), y	7	8.25	9.5	10.75

0.25	7	15
1	8.25	20
3.25	9.5	30

slope: $\dfrac{[\quad] - [\quad]}{[\quad] - [\quad]} = \dfrac{[\quad]}{[\quad]}$

y-intercept: [　　] = [　　]([　　]) + b; b = [　　]

linear function: y = [　　]x + [　　]

What do the slope and y-intercept of the function represent?

[　　　　　　　　　　　　　　　　　　　　　　　　　]

CCSS Common Core Spiral Review

Graph and label each ordered pair on a coordinate plane. 6.NS.6

15. A(3, 3)

16. D(1, 8)

17. G(2.5, 7)

18. X(7, 2)

19. P(0, 6)

20. N(4½, 0)

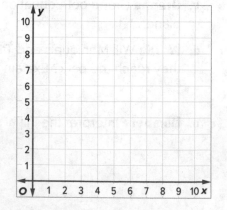

Linear and Nonlinear Functions

  Watch ▶

Football The table shows the approximate height and horizontal distance traveled by a football kicked at an angle of 30° with an initial velocity of 30 yards per second.

1. Is the rate of change for the height of the football constant? Explain.

_____

_____

_____

2. Is the rate of change for the distance traveled constant? Explain.

_____

_____

Time (s)	Height (yd)	Distance (yd)
0.0	0	0
0.5	6.2	13
1.0	9.7	26
1.5	10.5	39
2.0	8.7	52
2.5	4.2	65

 Essential Question

HOW can we model relationships between quantities?

Vocab a_b_c **Vocabulary**

nonlinear function

 CCSS **Common Core State Standards**

Content Standards
8.F.1, 8.F.3, 8.F.5
 MP **Mathematical Practices**
1, 3, 4, 7

3. Graph the ordered pairs (time, height) and (time, distance) on separate grids. Connect the points with a straight line or smooth curve. Then compare the graphs.

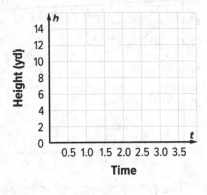

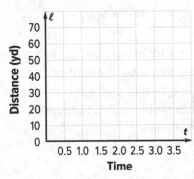

Which MP **Mathematical Practices did you use?**
Shade the circle(s) that applies.

① Persevere with Problems

② Reason Abstractly

③ Construct an Argument

④ Model with Mathematics

⑤ Use Math Tools

⑥ Attend to Precision

⑦ Make Use of Structure

⑧ Use Repeated Reasoning

Identify Linear and Nonlinear Functions

In a previous lesson, you learned that linear functions have graphs that are straight lines. This is because the rate of change between any two data points is a constant. **Nonlinear functions** are functions whose rates of change are not constant. Therefore, their graphs are not straight lines.

Examples

Determine whether each table represents a *linear* or *nonlinear* function. Explain.

1.

x	y
2	50
4	35
6	20
8	5

+2, +2, +2 on left; −15, −15, −15 on right

As x increases by 2, y decreases by 15 each time. The rate of change is constant, so this function is linear.

2.

x	y
1	1
4	16
7	49
10	100

+3, +3, +3 on left; +15, +33, +51 on right

As x increases by 3, y increases by a greater amount each time. The rate of change is not a constant, so this function is nonlinear.

Check

Graph the points on a coordinate plane.

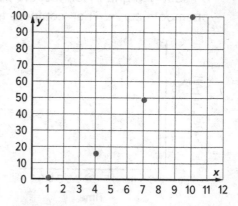

The points do not fall in a line. The function is nonlinear. ✓

Increasing and Decreasing Functions

If y increases as x increases, the function is called an increasing function. If y decreases as x increases, the function is called a decreasing function.

Show your work.

a. _____

b. _____

> **Got it?** Do these problems to find out.

Determine whether each table represents a *linear* or *nonlinear* function. Explain.

a.

x	0	5	10	15
y	20	16	12	8

b.

x	0	2	4	6
y	0	2	8	18

Example

3. Use the table to determine whether the minimum number of Calories a tiger cub should eat is a linear function of its age in weeks.

Age (weeks)	Minimum Calorie Intake
1	825
2	1,000
3	1,185
4	1,320
5	1,420

Use the table to find the rates of change.

$$1,000 - 825 = 175$$
$$1,185 - 1,000 = 185$$
$$1,320 - 1,185 = 135$$
$$1,420 - 1,320 = 100$$

The rates of change are not the same. Therefore, this function is nonlinear.

Check Graph the data to verify the ordered pairs do not lie on a straight line.

Got it? Do this problem to find out.

c. Tickets to the school dance cost $5 per student. Are the ticket sales a linear function of the number of tickets sold? Explain.

Number of Tickets Sold	1	2	3
Ticket Sales	$5	$10	$15

c. _____

Example

4. A square has a side length of *s* inches. The area of the square is a function of the side length. Does this situation represent a linear or nonlinear function? Explain.

Make a table to show the area of the square for side lengths of 1, 2, 3, 4, and 5 inches.

Side Length (in.)	1	2	3	4	5
Area (in²)	1	4	9	16	25

Graph the function. The function is not linear because the points (1, 1), (2, 4), (3, 9), (4, 16), and (5, 25) are not on a straight line.

d. _____

d. A square has a side length of *s* inches. The perimeter of the square is a function of the side length. Does this situation represent a linear or nonlinear function? Explain.

Guided Practice

Determine whether each table represents a *linear* or *nonlinear* function. Explain. (Examples 1 and 2)

1.

x	0	1	2	3
y	1	3	6	10

2.

x	0	3	6	9
y	−3	9	21	33

 Show your work.

_____

_____

3. The table shows the measures of the sides of several rectangles. Are the widths of the rectangles a linear function of the lengths? Explain. (Example 3) _____

Length (in.)	1	4	8	10
Width (in.)	64	16	8	6.4

_____

4. A cube has a side length of *s* meters. The volume of the cube is represented by the expression s^3. The volume of the cube is a function of the side length. Does this situation represent a linear or nonlinear function? Explain. (Example 4) _____

_____

_____

5. **Building on the Essential Question** How can you use a table or a graph to determine if a function is linear or nonlinear?

_____

_____

_____

_____

_____

Rate Yourself!

How confident are you about functions? Check the box that applies.

For more help, go online to access a Personal Tutor.

 Tutor

 FOLDABLES Time to update your Foldable!

Independent Practice

Go online for Step-by-Step Solutions

Determine whether each table represents a *linear* or *nonlinear* function. Explain. (Examples 1 and 2)

1

x	−2	0	2	4
y	−1	0	1	2

Show your work.
_____
_____

2.

x	1	2	3	4
y	1	4	9	16

_____
_____

3.

x	5	10	15	20
y	13	28	43	58

_____
_____

4.

x	1	3	5	7
y	−2	−18	−50	−98

_____
_____

5 The Guzman family drove from Anderson to Myrtle Beach. Use the table to determine whether the distance driven is a linear function of the hours traveled. Explain. (Example 3) _____
_____

Time (h)	1	2	3	4
Distance (mi)	65	130	195	260

6. The table shows the height of several buildings in Chicago. Use the table to determine whether the height of the building is a linear function of the number of stories. Explain. (Example 3)

_____
_____

Building	Stories	Height (ft)
Harris Bank III	35	510
One Financial Place	40	515
Kluczynski Federal Building	45	545
Mid Continental Plaza	50	582
North Harbor Tower	55	552

7. There are 3,600 seconds in one hour. The total seconds is a function of the hours. Does this situation represent a linear or nonlinear function? Explain. (Example 4)

_____
_____

8. A football is placed on the ground to kick a field goal. The height of the ball is a function of the time in seconds. Does the path the football follows after being kicked represent a linear or nonlinear function? Explain. (Example 4)

_____
_____

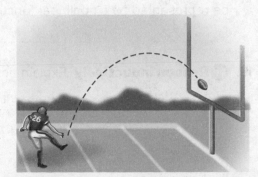

Graph each function by making a table of ordered pairs. Determine whether each function is *linear* or *nonlinear*. Explain.

9. $y = -x + 1$

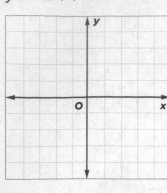

_____

_____

10. $y = \dfrac{-4}{x}$

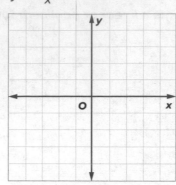

_____

_____

11. $y = \dfrac{3x}{2}$

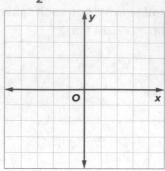

_____

_____

12. **MP Identify Structure** Complete the graphic organizer by determining if the graphs represent linear or nonlinear functions.

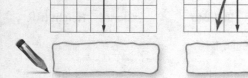

Linear or Nonlinear?

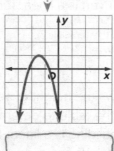

🔥 H.O.T. Problems Higher Order Thinking

13. **MP Persevere with Problems** Does the graph at the right represent a linear function? Explain. _____

_____

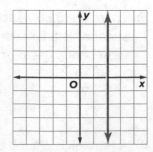

14. **MP Model with Mathematics** Give an example of a situation that can be represented by a nonlinear function. _____

_____

15. **MP Reason Inductively** Explain how you can use different representations to determine whether a function is linear. _____

_____

Extra Practice

Determine whether each table represents a *linear* or *nonlinear* function. Explain.

16.

x	y
2	10
4	12
6	16
8	24

+2 → +2
+2 → +4
+2 → +8

Homework Help

Nonlinear; rate of change is not constant. As x increases by 2, y increases by a greater amount each time. The rate of change is not constant, so this function is nonlinear.

17.

x	y
4	3
8	0
12	−3
16	−6

_____

_____

_____

_____

18. Copy and Solve The area of a square is a function of its perimeter. Graph the function on a separate sheet of grid paper. Explain whether the function is linear and if the graph is increasing or decreasing.

19. MP Multiple Representations Recall that the circumference of a circle is equal to two times π times its radius and that the area of a circle is equal to π times the square of the radius.

a. Tables Complete the table showing the circumference and area of circles with radius *r*.

Radius r	Circumference 2 • π • r	Area πr²
1	2 • π • 1 ≈ 6.28	π • 1² ≈ 3.14
2		
3		
4		
5		

b. Graphs Graph the ordered pairs (radius, circumference) and (radius, area) for each function on the same coordinate plane.

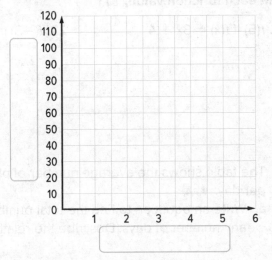

c. Words Is the circumference of a circle a linear or nonlinear function of its radius? the area?

Explain your reasoning. _____

_____

20. Determine if each table represents a linear or nonlinear function.

linear
nonlinear

x	y
9	1.25
11.5	2
14	2.75
16.5	3.5

x	y
−3	−6
−7	−1
−11	4
−15	9

x	y
10	5
13	7
16	10
19	14

x	y
2	15
4	20
6	25
8	30

21. Jung has $200 in a safe. Each month, he adds another $10 to the safe. Miguel opens a savings account with a $200 deposit and earns 2.5% interest each month on the total amount of money in the bank. Determine if each statement is true or false.

a. The function representing Jung's savings is nonlinear. ☐ True ☐ False

b. The function representing Miguel's savings is linear. ☐ True ☐ False

c. After 1 year, Jung will have saved $320. ☐ True ☐ False

CCSS **Common Core Spiral Review**

Find each function value. 8.F.1

22. $f(5)$ if $f(x) = 3x + 4$ _____

23. $f(-3)$ if $f(x) = 2x - 8$ _____

24. $f(7)$ if $f(x) = 9x - 24$ _____

25. The table shows the average number of phone calls Riley makes per day. 8.F.4

a. Write an equation to find the total number of phone calls made in any number of days. Describe the relationship in words.

b. Use the equation to determine how many phone calls Riley would make in 1 week.

Number of Days, d	Total Phone Calls, c
1	5
2	10
3	15
4	20

Quadratic Functions

Vocabulary Start-Up

In the previous lesson, you learned about nonlinear functions. A special type of a nonlinear function is a quadratic function. A **quadratic function** is a function in which the greatest power of the variable is 2. Its graph is U-shaped, opening upward or downward.

Complete the graphic organizer by determining if the facts about quadratic functions are *true* or *false*. If false, give the true fact.

A function in which the greatest power of the variable is 2.

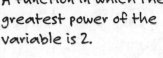

Quadratic Functions

The graph is a straight line.

The graph of a quadratic function sometimes opens upward.

The graph always opens downward.

 Essential Question

HOW can we model relationships between quantitites?

 Vocabulary

quadratic function

Common Core State Standards

Content Standards
8.F.3, 8.F.5

MP Mathematical Practices
1, 3, 4, 7

 ## Real-World Link

Kris kicked a soccer ball straight into the air. The height h in feet of the ball after t seconds is found using the equation $h = -16t^2 + 40t + 2$. What is the height of the ball after 1.5 seconds? _____

Which MP Mathematical Practices did you use?
Shade the circle(s) that applies.

① Persevere with Problems
② Reason Abstractly
③ Construct an Argument
④ Model with Mathematics

⑤ Use Math Tools
⑥ Attend to Precision
⑦ Make Use of Structure
⑧ Use Repeated Reasoning

Quadratic Functions

A quadratic function can be written in the form $y = ax^2 + bx + c$, where $a \neq 0$. The graph of a quadratic function is called a parabola. The graph opens upward if the coefficient of the variable that is squared is positive, downward if it is negative.

Examples

1. Graph $y = x^2$.

To graph a quadratic function, make a table of values, plot the ordered pairs, and connect the points with a smooth curve.

x	x^2	y	(x, y)
-2	$(-2)^2 = 4$	4	$(-2, 4)$
-1	$(-1)^2 = 1$	1	$(-1, 1)$
0	$(0)^2 = 0$	0	$(0, 0)$
1	$(1)^2 = 1$	1	$(1, 1)$
2	$(2)^2 = 4$	4	$(2, 4)$

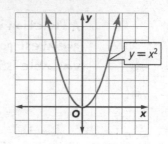

2. Graph $y = -x^2 + 4$.

x	$-x^2 + 4$	y	(x, y)
-2	$-(-2)^2 + 4 = 0$	0	$(-2, 0)$
-1	$-(-1)^2 + 4 = 3$	3	$(-1, 3)$
0	$-(0)^2 + 4 = 4$	4	$(0, 4)$
1	$-(1)^2 + 4 = 3$	3	$(1, 3)$
2	$-(2)^2 + 4 = 0$	0	$(2, 0)$

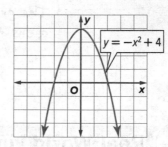

Got it? Do this problem to find out.

a. Graph $y = 6x^2$.

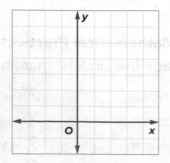

Examples

3. The function $d = 4t^2$ represents the distance d in feet that a race car will travel over t seconds with a constant acceleration of 8 ft/s^2. Graph the function. Then use the graph to find how much time it will take for the race car to travel 200 feet.

Time cannot be negative, so only use positive values of t.

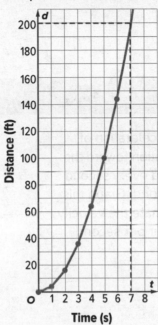

t	$d = 4t^2$	(t, d)
0	$4(0)^2 = 0$	$(0, 0)$
1	$4(1)^2 = 4$	$(1, 4)$
2	$4(2)^2 = 16$	$(2, 16)$
3	$4(3)^2 = 36$	$(3, 36)$
4	$4(4)^2 = 64$	$(4, 64)$
5	$4(5)^2 = 100$	$(5, 100)$
6	$4(6)^2 = 144$	$(6, 144)$

Locate 200 on the vertical axis. Move over to the graph and locate the corresponding value for the time.

The car will travel 200 feet after about 7 seconds.

4. The function $h = 0.66d^2$ represents the distance d in miles you can see from a height of h feet. Graph this function. Then use the graph to estimate how far you could see from a hot air balloon 1,000 feet in the air.

Distance cannot be negative, so use only positive values of d.

d	$h = 0.66d^2$	(d, h)
0	$0.66(0)^2 = 0$	$(0, 0)$
10	$0.66(10)^2 = 66$	$(10, 66)$
20	$0.66(20)^2 = 264$	$(20, 264)$
25	$0.66(25)^2 = 412.5$	$(25, 412.5)$
30	$0.66(30)^2 = 594$	$(30, 594)$
35	$0.66(35)^2 = 808.5$	$(35, 808.5)$
40	$0.66(40)^2 = 1,056$	$(40, 1,056)$

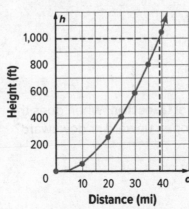

At a height of 1,000 feet, you could see approximately 39 miles.

Got it? Do this problem to find out.

b. _____

b. The outdoor observation deck of the Space Needle in Seattle, Washington, is 520 feet above ground level. Use the graph to estimate how far you could see from the observation deck.

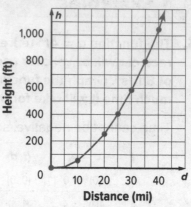

Guided Practice

1. The function $a = 0.2v^2$ models the acceleration of a carnival ride, where a is the acceleration toward the center of the ride in meters per second every second and v is the velocity in meters per second. Graph this function. Then use your graph to estimate the velocity of the ride at an acceleration of 1 meter per second

every second. (Examples 3 and 4) _____

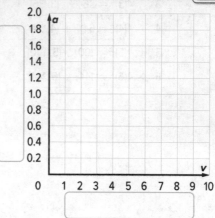

Graph each function. (Examples 1 and 2)

2. $y = 3x^2$

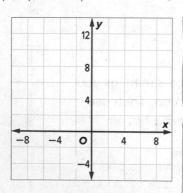

3. $y = -5x^2$

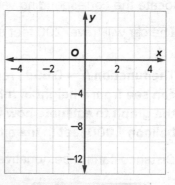

4. **Building on the Essential Question** When does the graph of a quadratic function open upward or downward?

_____

_____

_____

Rate Yourself!

How confident are you about quadratic functions? Check the box that applies.

For more help, go online to access a Personal Tutor.

 Tutor

Independent Practice

Go online for Step-by-Step Solutions

eHelp

Graph each function. (Examples 1 and 2)

1 $y = 4x^2$

Show your work.

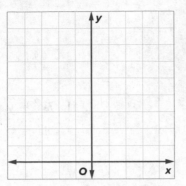

2. $y = -3x^2$

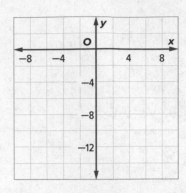

3. A penny is dropped from a height of 196 feet off a bridge. The function $d = -16t^2 + 196$ models the distance d in feet the penny is from the surface of the water at time t seconds. Graph this function. Then use your graph to estimate the time it will take for the penny to reach the water. (Examples 3 and 4) _____

4. The area A in square feet of a projected movie on a movie screen can be represented by the equation $A = 0.25d^2$, where d represents the distance from a projector to the movie screen. Graph the function. Then use your graph to estimate the distance from the projector to a screen if the area of the movie is 7 square feet.

(Examples 3 and 4) _____

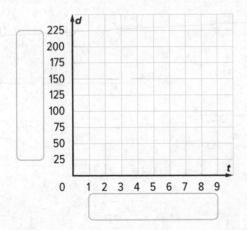

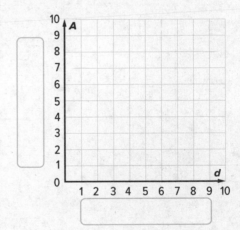

5. Anna has trim to make a rectangular border for a scrapbook page. The section inside the border is x inches long and $(12 - x)$ inches wide.

  a. Write a function to represent the area A of the section inside the border. _____

  b. What should the dimensions of the section be to enclose the maximum area inside the border? (*Hint*: Graph the function and find the x-coordinate of the point at the peak of the graph.) _____

MP Identify Structure Without graphing, determine whether each equation represents a linear or nonlinear function. Explain.

6. $y = 3x$

7. $y = 2x^2$

8. $y = -3x^2$

9. $y = -6x$

10. $5x + y = 7$

11. $7x^2 + y = 24$

H.O.T. Problems Higher Order Thinking

12. MP Persevere with Problems The graphs of quadratic functions may have exactly one highest point, called a *maximum*, or exactly one lowest point, called a *minimum*. Graph each quadratic equation. Determine whether each graph has a maximum or a minimum. If so, give the coordinates of each point.

a. $y = 2x^2 + 1$

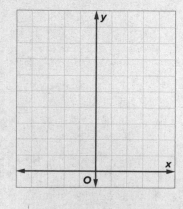

b. $y = -x^2 + 5$

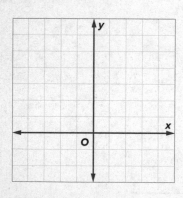

c. $y = x^2 - 3$

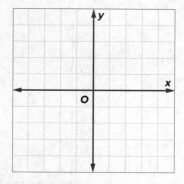

13. MP Model with Mathematics Write the equation of a quadratic function that opens upward and has its minimum at $(0, -3.5)$.

14. MP Reason Inductively The equation $y = ax^2 + bx + c$ represents a quadratic function. What does the constant c represent? Explain.

Extra Practice

Graph each function.

15. $y = -x^2 + 2$

x	$-x^2 + 2$	y	(x, y)
-2	$-(-2)^2 + 2$	-2	(-2, -2)
-1	$-(-1)^2 + 2$	1	(-1, 1)
0	$-(0)^2 + 2$	2	(0, 2)
1	$-(1) + 2$	1	(1, 1)
2	$-(2)^2 + 2$	-2	(2, -2)

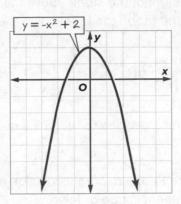

16. $y = -x^2 - 4$

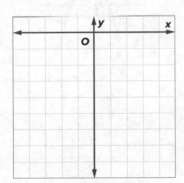

17. $y = 2x^2 + 3$

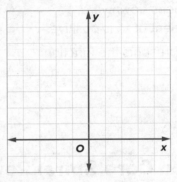

18. The function $d = 0.006s^2$ represents the braking distance d in meters of a car traveling at a speed s in kilometers per second. Graph this function. Then use your graph to estimate the speed of the car if its braking distance is 12 meters. _____

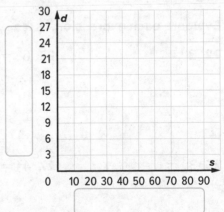

19. **MP Model with Mathematics** Annika is making a fabric memo board. The width of the board is x inches, and the length of the board is $(x + 4)$ inches.

a. Write a function that represents the area A of the memo board. _____

b. Graph the function.

c. If the width of the memo board is 8 inches, what is its area? _____

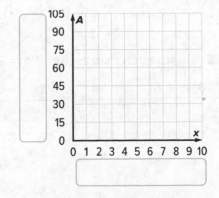

20. The height of a stone that is dropped from a 144-foot tall bridge can be modeled by the function $h = -16t^2 + 144$, where t is the time in seconds and h is the height of the stone above the river. Complete the table of values below and graph the function on the coordinate plane.

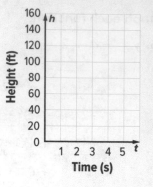

Time, t	0	0.5	1	1.5	2	2.5	3
Height, h							

How long does it take for the stone to reach the river?

21. Rachel sketched the graph of a quadratic function as shown. Determine if each statement is true or false.

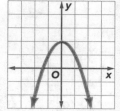

a. The y-intercept is (0, 2). ☐ True ☐ False

b. The graph opens downward, so the coefficient of x^2 is positive. ☐ True ☐ False

c. The graph represents the function $y = -x^2 + 2$. ☐ True ☐ False

Common Core Spiral Review

Evaluate each expression. 6.EE.2

22. $2^4 =$ _____

23. $6^4 =$ _____

24. $8^3 =$ _____

25. $3^5 =$ _____

26. $3^3 =$ _____

27. $4^2 =$ _____

28. $5^4 =$ _____

29. $6^2 =$ _____

Evaluate each expression if $a = 2$, $b = -3$, and $c = 6$. 7.NS.3

30. $a + b - c$ _____

31. $c - a + b$ _____

32. $a \times c$ _____

33. $\dfrac{c}{a}$ _____

Inquiry Lab

Graphing Technology: Families of Nonlinear Functions

 HOW are families of nonlinear functions the same as the parent function? How are families of nonlinear functions different from the parent function?

 Content Standards 8.F.3, 8.F.5

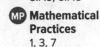 **Mathematical Practices** 1, 3, 7

Sasha is digging a vegetable garden and wants to know how much fertilizer to buy. The equation $y = x^2$ represents the area in square feet of the garden shown.

Hands-On Activity 1

Families of nonlinear functions share a common characteristic based on a parent function. The parent function, or simplest function, of a family of quadratic functions is $y = x^2$. You can use a graphing calculator to investigate families of quadratic functions.

Graph $y = x^2$, $y = x^2 + 5$, and $y = x^2 - 3$ on the same screen.

Step 1 Clear any existing equations from the Y= list by pressing [Y=] [CLEAR].

Step 2 Enter each equation. Press [X,T,θ,n] [x²] [ENTER], [X,T,θ,n] [x²] [+] 5 [ENTER], and [X,T,θ,n] [x²] [−] 3 [ENTER].

Step 3 Press [ZOOM] 6.

Copy your calculator screen on the blank screen shown.

How are the three equations related?

_____

_____

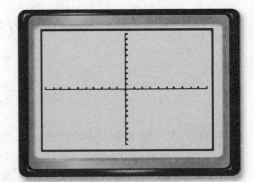

Describe how the graphs of the three equations are related.

_____

_____

_____

Hands-On Activity 2

An **exponential function** is a nonlinear function in which the base is a constant and the exponent is an independent variable, x. The parent function for an exponential function is shown.

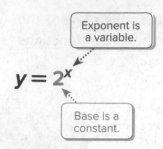

Exponent is a variable.

$$y = 2^x$$

Base is a constant.

A certain type of bacteria doubles every hour. The function $y = 2^x$ represents the total number of bacteria y at the end of every hour x. Graph the function. Then find the number of bacteria at the end of 5 hours.

Step 1 Clear any existing equations from the Y= list by pressing
Y= CLEAR .

Step 2 Enter the equation. Press Y= 2 ∧ X,T,θ,n

Step 3 Graph the equation in the standard viewing window. Press ZOOM 6.

Copy your calculator screen on the blank screen shown.

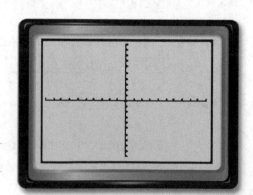

Describe the function by analyzing the graph.

_____

_____

Step 4 Use the TABLE feature. Press 2nd GRAPH . What is the y value that corresponds to the x value of 5? ⬚

So, there are ⬚ bacteria at the end of 5 hours.

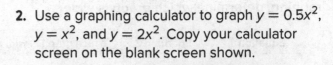

Investigate

Work with a partner.

1. Use a graphing calculator to graph $y = x^2$, $y = x^2 - 6$, and $y = x^2 + 4$. Copy your calculator screen on the blank screen shown.

How does changing the value of c in the equation $y = x^2 + c$ affect the graph?

_____

_____

_____

2. Use a graphing calculator to graph $y = 0.5x^2$, $y = x^2$, and $y = 2x^2$. Copy your calculator screen on the blank screen shown.

How does changing the value of a in the equation $y = ax^2$ affect the graph?

_____

_____

_____

3. Use a graphing calculator to graph $y = 2^x$, $y = 2^x + 1$, and $y = 2^x - 3$. Copy your calculator screen on the blank screen shown.

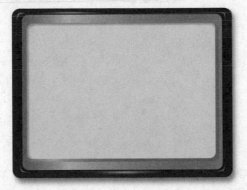

How does changing the value of c in the equation $y = 2^x + c$ affect the graph?

_____

_____

_____

4. Use a graphing calculator to graph $y = 0.5^x$, $y = 0.25^x$ and $y = 2^x$. Copy your calculator screen on the blank screen shown.

How does changing the value of a to a fraction in the equation $y = a^x$ affect the graph?

_____

_____

_____

Analyze and Reflect

MP Identify Structure Work with a partner to complete the table. Without graphing, determine which graph is wider.

	Equation 1	Equation 2	Which graph is wider?
5.	$y = 5x^2$	$y = x^2$	
6.	$y = \frac{1}{3}x^2$	$y = 3x^2$	
7.	$y = 2^x$	$y = 4^x$	
8.	$y = 0.25^x$	$y = 0.75^x$	

9. Miley's parents started a savings account when she was born by depositing $100 into the account. The account has an annual interest rate of 3%. The balance y in the account can be represented by the function $y = 100(1.03)^x$, where x is the number of years.

   a. Graph the function on your graphing calculator. Copy your calculator screen on the blank screen shown. (*Hint:* Use the x-scale −50 to 50 and the y-scale 0 to 500.)

   b. Use the TABLE feature. How much money is in the

   account after 13 years? _____

   c. Describe the function by analyzing the graph.

   _____

   _____

Create

10. **MP Model with Mathematics** Write the equation of a quadratic function that is wider than $y = \frac{2}{3}x^2$. Explain how you know it is wider.

   _____

   _____

11. **Inquiry** HOW are families of nonlinear functions the same as the parent function? How are families of nonlinear functions different from the parent function?

   _____

   _____

   _____

Qualitative Graphs

Downloads Emily is downloading photos from her digital camera to her computer. The table shows the percent of photos downloaded for several seconds.

Time (s)	Percent Downloaded
0	0
2	15
4	30
6	30
8	64
10	64
12	82
14	100

1. During which period(s) of time did the percent downloaded not change?

_____

_____

2. During which period of time did the percent downloaded change the most?

_____

3. Graph and connect the ordered pairs.

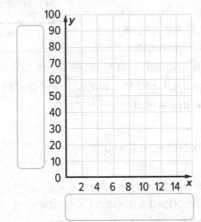

 Essential Question

HOW can we model relationships between quantities?

 Vocab

Vocabulary

qualitative graphs

 Common Core State Standards

Content Standards
8.F.5

MP **Mathematical Practices**
1, 2, 3, 4

Which **MP** Mathematical Practices did you use?
Shade the circle(s) that applies.

① Persevere with Problems

② Reason Abstractly

③ Construct an Argument

④ Model with Mathematics

⑤ Use Math Tools

⑥ Attend to Precision

⑦ Make Use of Structure

⑧ Use Repeated Reasoning

Analyze Qualitative Graphs

The graph shown is a qualitative graph. **Qualitative graphs** are graphs used to represent situations that may not have numerical values or graphs in which numerical values are not included.

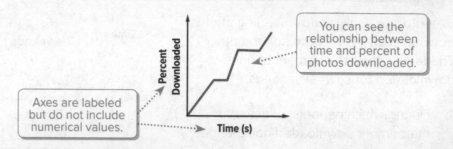

You can see the relationship between time and percent of photos downloaded.

Axes are labeled but do not include numerical values.

Example

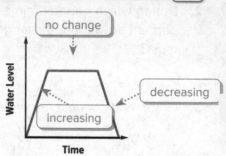

1. **The graph at the right displays the water level in a bathtub. Describe the change in the water level over time.**

At time zero, the water level in the bathtub is zero. The water level in the bathtub increases at a constant rate. Then the water is turned off and the water level does not change. Finally, the drain plug is pulled and the water level decreases at a constant rate until the water level is zero.

Got it? Do these problems to find out.

a. The graph at the right displays the temperature throughout the day. Describe the change in the temperature over time.

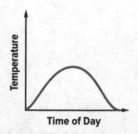

b. The graph represents revenue from a local clothing store. Describe the sales over time.

a. _____

b. _____

Show your work.

Sketch Qualitative Graphs

Qualitative graphs represent the essential elements of a situation in a graphical form. You can sketch qualitative graphs to represent many real-world functions that are described verbally.

 Real World

Examples Tutor

2. **A tennis ball is dropped onto the floor. On each successive bounce, it rebounds to a height less than its previous bounce height until it comes to rest on the floor. Sketch a qualitative graph to represent the situation.**

Step 1 Draw the axes. Label the vertical axis "Distance from Floor." Label the horizontal axis "Time."

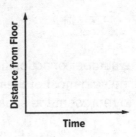

Step 2 Sketch the shape of the graph. The distance from the floor starts out at a high value. The ball falls to the floor, bounces, and rebounds to a height less than its drop height. This pattern is repeated several times until the ball comes to rest on the floor.

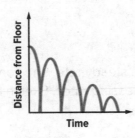

3. **A child swings on a swing. Sketch a qualitative graph to represent the situation.**

Step 1 Draw the axes. Label the vertical axis "Distance from the ground" and the horizontal axes "Time Elapsed."

Step 2 Sketch the shape of the graph. The distance from the swing to the ground starts at a low value. The child continues to swing and creates momentum each time the swing goes back until the child on the swing stops.

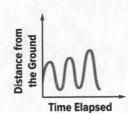

Got it? Do this problem to find out.

c. A car is traveling at a constant speed. The car slows down steadily to come to rest at a stop light. Sketch a qualitative graph to represent the situation.

c. _____

Guided Practice

1. The graph at the right displays the height of an airplane. Describe the change in the airplane's height over time. (Example 1)

_____

_____

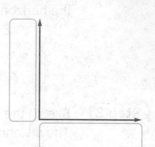

2. Jamaal purchased the same number of action figures daily for one week. Over the next week, he sold most of them on the Internet. Sketch a qualitative graph to represent the situation. (Examples 2 and 3)

3. Tamar rides her bicycle at a steady rate. She coasts downhill which increases her speed at increasing rates. Sketch a qualitative graph to represent the situation. (Examples 2 and 3)

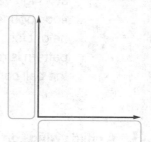

4. 🅠 **Building on the Essential Question** What are some advantages of displaying the relationship between two quantities using a qualitative graph?

_____

_____

_____

_____

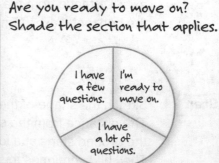

Rate Yourself!

Are you ready to move on?
Shade the section that applies.

I have a few questions.

I'm ready to move on.

I have a lot of questions.

For more help, go online to access a Personal Tutor.

Tutor

Independent Practice

Go online for Step-by-Step Solutions

1 The graph below displays the distance from Luis' home as he walks his dog in his neighborhood. Describe the change in the distance from his home over time. (Example 1)

 Show your work.

2. The graph below displays the speed of a city bus as it stops frequently to pick up passengers. Describe the change in the speed over time. (Example 1)

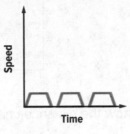

_____

_____

_____

3 A grand piano that is over 100 years old has increased in value rapidly from when it was first purchased. Sketch a qualitative graph to represent this situation. (Examples 2 and 3)

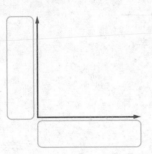

4. An athlete alternates between running and walking during a workout. Sketch a qualitative graph to represent this situation. (Examples 2 and 3)

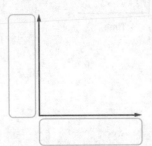

5. **MP Reason Abstractly** Use the graph at the right which displays the rate at which Hector hiked along a path.

a. What situation could the horizontal line segment represent?

_____

b. What situation could the vertical line segment represent?

_____

c. Did Hector's rate increase or decrease during the first portion of his hike? Explain your reasoning.

_____

_____

6. **MP** **Persevere with Problems** The graph at the right displays the speed of a car as time increases.

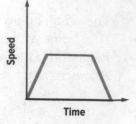

Speed / Time

a. Draw a qualitative graph that represets the distance the car travels as time increases.

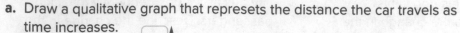

b. Describe how the distance changes as time passes.

_____

_____

_____

7. **MP** **Reason Inductively** A tree grows steadily. When it reaches a specific height, it stops growing. Which graph displays this relationship? Explain your reasoning to a classmate.

Graph A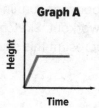

Graph B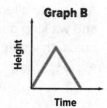

Graph C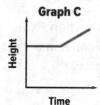

_____

_____

_____

8. **MP** **Reason Inductively** The graph below represents the height of a rocket after being launched.

Describe the change in the height of the rocket over time.

_____

_____

_____

9. The graph below displays the distance Justine rode on her bike. Describe the change in the distance over time.

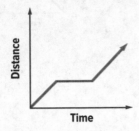

Justine rode her bike at a constant rate in the beginning. She then stopped riding for a period of time. Then she continued riding at a constant rate.

10. The graph below displays the temperature of a cup of hot chocolate. Describe the change in the temperature over time.

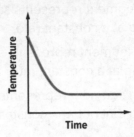

_____

_____

_____

11. A graph of Mrs. Fraser's electric bill throughout the year is shown below. Describe the change in the bill over time.

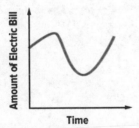

_____

_____

_____

_____

12. The graph below displays the distance covered on a long road trip. Describe the change in the distance over time.

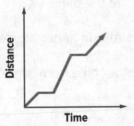

_____

_____

_____

_____

13. **MP Model with Mathematics** The outside temperature rises throughout the day at varied rates, then drops at night. Sketch a qualitative graph to represent the situation.

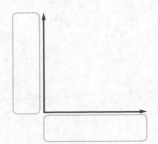

14. A lion cub is resting in the grass. He sees another lion cub nearby and races after it, picking up speed as it runs. Sketch a qualitative graph to represent the situation.

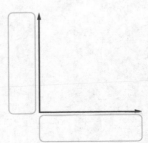

15. The graph represents the amount of pies sold by a bakery over the course of one day. Determine if each of the following statements is true or false.

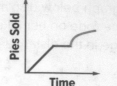

   a. The red segment represents sales increasing at a constant rate. ☐ True ☐ False

   b. The blue segment represents sales, decreasing at a constant rate. ☐ True ☐ False

   c. The green segment represents sales increasing, but not at a constant rate. ☐ True ☐ False

16. The graph represents Jenny's activities on her way home from school on a given day. Match each statement with its corresponding segment on the graph.

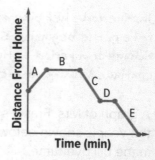

Jenny rides her bike from the park to Allison's house. ☐

Jenny rides her bike home from Allison's house. ☐

Jenny plays at the park. ☐

Jenny visits Allison at her house. ☐

Jenny rides her bike from school to the park. ☐

Common Core Spiral Review

Simplify each expression. 7.EE.1

17. $2(p + 8) + 4 =$ _____

18. $(18 + t)(-3) + 9 =$ _____

19. $30q(2) =$ _____

20. $-5(n + 16) - 7 =$ _____

21ST CENTURY CAREER
in Physical Therapy

Physical Therapist

Are you a compassionate person? Do you have a strong desire to help others? If so, a career as a physical therapist might be a good choice for you. Physical therapists help restore function, improve mobility, and relieve pain of patients suffering from injuries or disease. One of their jobs is to teach exercises or recommend activities to help patients regain balance, flexibility, endurance, and strength.

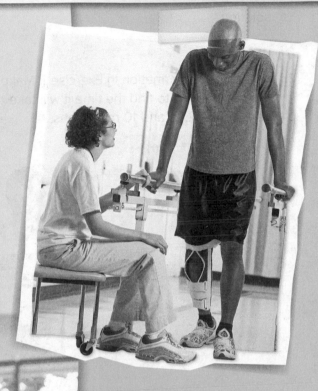

Is This the Career for You?

Are you interested in a career as a physical therapist? Take some of the following courses in high school.

◆ Algebra
◆ Biology
◆ Chemistry
◆ Introduction To Physical Therapy

Turn the page to find out how math relates to a career in Physical Therapy.

Focusing on Recovery

Use the information in the table below to solve each problem.

1. The function $t(r) = 12r$, where r is the number of repetitions, represents the total time $t(r)$ in seconds to complete a flexibility exercise. Find $t(8)$. Then interpret the solution. _____

2. Refer to the information in Exercise 1. Make a function table to find the time it will take to complete 1, 2, 5, and 10 repetitions.

r	12r	t(r)

3. Write a function to represent the distance d in miles a runner will travel in t minutes.

_____

4. Refer to the function that you wrote in Exercise 3. How far will a runner travel after 80 minutes? _____

5. Graph the function from Exercise 3. Then use the graph to estimate the distance a runner will travel after 90 minutes.

_____

Endurance Exercise: Cross-Country Running

Time (min)	Distance (mi)
15	2.25
30	4.5
45	6.75
60	9.0

Career Project

It's time to update your career portfolio! Make a list of questions that you would like to know about a career in physical therapy. Then interview a physical therapist in your area. Include all the interview questions and answers in your portfolio.

_____

_____

_____

_____

List other careers that someone with an interest in physical therapy could pursue.

- _____
- _____
- _____
- _____
- _____

Chapter Review

Vocabulary Check

Reconstruct the vocabulary word and definition from the letters under the grid. The letters for each column are scrambled directly under that column.

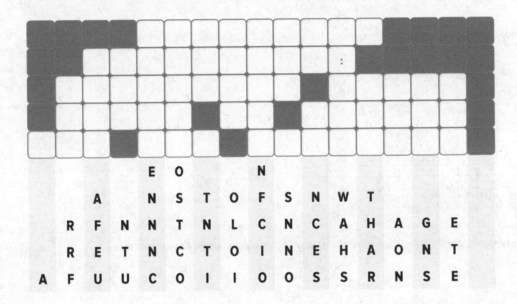

```
                E   O           N
            A   N   S   T   O   F   S   N   W   T
        R   F   N   N   T   N   L   C   N   C   A   H   A   G   E
        R   E   T   N   C   T   O   I   N   E   H   A   O   N   T
    A   F   U   U   C   O   I   I   O   O   S   S   R   N   S   E
```

Complete each sentence using the vocabulary list at the beginning of the chapter.

1. A _____ is any set of ordered pairs.

2. The variable for the range is called the _____
   because it depends on the domain.

3. Graphs used to represent situations that may not have numerical
   values or graphs in which numerical values are not included are

   called _____.

4. The variable for the domain is called the _____
   because it can be any number.

5. _____ can take on any value, so there is
   no space between data values for a given domain.

6. A function in which the greatest power of the variable is 2 is called

   a _____.

7. A _____ is a relation in which every member of the domain
   (input values) is paired with exactly one member of the range (output value).

Use Your FOLDABLES

Use your Foldable to help review the chapter.

Tape here

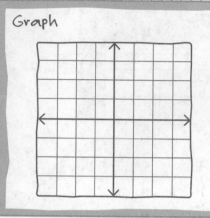

Tab 3   Relations and Functions

Tab 2

Tab 1

Graph

Graph

Got it?

Match each equation with the correct graph.

1. $-x + y = -3$

2. $2x + y = 6$

3. $y = 2x^2 + 1$

4. $y = -x^2 + 2$

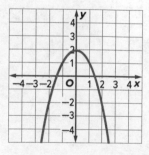

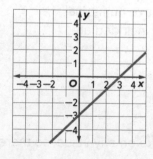

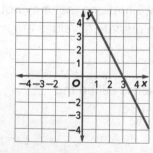

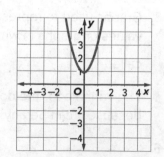

Power Up! Performance Task

Tournament Time

Lake City High School is hosting a middle school basketball tournament. The number of teams involved, x, is still unknown. However, the total number of games, $g(x)$, is related to the number of teams in the tournament. The relationship between the possible number of teams and the total number of games played is shown in the table.

Teams, x	2	3	4	5	6
Games, $g(x)$	1	2	3	4	5

Write your answers on another piece of paper. Show all of your work to receive full credit.

Part A
Write a function that relates the number of teams to the total number of games played.

Part B
The function $r(x) = 50x - 50$ represents the cost to have two referees for each game played. What is the cost for referees if there are 2, 3, 4, 5, and 6 teams in the tournament?

Part C
Graph the ordered pairs for the function in Part B on a coordinate plane. Is $r(x)$ a linear function? Explain.

Part D
The school wants to spend no more than $750 on referee fees. How many teams can enter the tournament?

Reflect

 Answering the Essential Question

Use what you learned about functions to complete the graphic organizer.
List four ways functions are expressed in this chapter.

Essential Question

HOW can we model relationships between quantities?

Answer the Essential Question. HOW can we model relationships between quantities?

_____

_____

UNIT PROJECT

Green Thumb If you have a knack for gardening, volunteering in a community garden is a great way to get involved with your community and also earn a little money. In this project you will:

- **Collaborate** with your classmates as you research the costs involved with growing vegetables and predict possible profits.
- **Share** the results of your research in a creative way.
- **Reflect** on how you find and use patterns to model real-world situations.

By the end of this Project, you just might be a young entrepreneur!

Collaborate

Go Online Work with your group to research and complete each activity. You will use your results in the Share section on the following page.

1. Choose a vegetable that is sold individually, and find its cost at a grocery store. Write an equation to represent the total cost as a function of the number of vegetables. Make a function table to find the cost of 1, 2, 3, 4, 5, and 6 vegetables. Then graph the ordered pairs.

2. Research a vegetable you would like to grow in a community garden. Find the costs involved such as buying seeds and gardening tools. Then determine how much you will charge per vegetable (or per pound) based on grocery store or farm market prices.

3. Based on the information you found in Exercise 2, write a linear function to represent your profit. Describe what the variables represent. Then graph and describe the function.

4. Research the following terms: *gross profit, total revenue, and gross profit margin*. Make a diagram explaining these terms. Then find your gross profit margin based on estimated gross profit and total revenue. What does your gross profit tell you?

5. Research the average temperatures in your area for the growing season of the vegetable you chose. Then sketch a qualitative graph that shows the change in temperature over the growing season. Include a brief explanation of your graph.

Share

With your group, decide on a way to share what you have learned about growing and selling vegetables. Some suggestions are listed below, but you can also think of other creative ways to present your information. Remember to show how you used mathematics to complete each of the activities in this project!

- Imagine you sell your vegetables at a farmer's market. Describe your experience in a blog.
- Use a budget spreadsheet to show how your vegetable can generate a profit. Include tables, equations, and graphs.

Check out the note on the right to connect this project with other subjects.

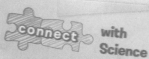

with Science

Environmental Literacy Research information about Earth's soil and the qualities needed to grow plants. Some questions to consider are:

- What type of soil allows fruits and vegetables to grow well?
- What type of soil is typically found in your area?
- What could you add to the soil to make it better for growing your plants?

Reflect

On Your Own

6. **Answer the Essential Question** How can you find and use patterns to model real-world situations?

   a. How did you use what you learned about constructing functions in this chapter to find and use patterns to model real-world situations in this project?

   _____

   _____

   _____

   b. How did you use what you learned about different representations of functions in this chapter to find and use patterns to model real-world situations in this project?

   _____

   _____

   _____

Glossary/Glosario

The eGlossary contains words and definitions in the following 13 languages:

Arabic
Bengali
Brazilian Portuguese

Cantonese
English
Haitian Creole

Hmong
Korean
Russian

Spanish
Tagalog

Urdu
Vietnamese

English	Español

Aa

accuracy The degree of closeness of a measurement to the true value.

acute angle An angle whose measure is less than 90°.

acute triangle A triangle with all acute angles.

Addition Property of Equality If you add the same number to each side of an equation, the two sides remain equal.

adjacent angles Angles that share a common vertex, a common side, and do not overlap. In the figure, the adjacent angles are ∠5 and ∠6.

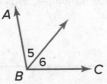

algebra A branch of mathematics that involves expressions with variables.

algebraic expression A combination of variables, numbers, and at least one operation.

exactitud Cercanía de una medida a su valor verdadero.

ángulo agudo Ángulo que mide menos de 90°.

triángulo acutángulo Triángulo con todos los ángulos agudos.

propiedad de adición de la igualdad Si sumas el mismo número a ambos lados de una ecuación, los dos lados permanecen iguales.

ángulos adyacentes Ángulos que comparten un vértice, un lado común y no se traslapan. En la figura, los ángulos adyacentes son ∠5 y ∠6.

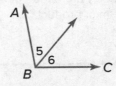

álgebra Rama de las matemáticas que trabaja con expresiones con variables.

expresión algebraica Una combinación de variables, números y por lo menos una operación.

alternate exterior angles Exterior angles that lie on opposite sides of the transversal. In the figure, transversal t intersects lines ℓ and m. ∠1 and ∠7, and ∠2 and ∠8 are alternate exterior angles. If line ℓ and m are parallel, then these pairs of angles are congruent.

ángulos alternos externos Ángulos externos que se encuentran en lados opuestos de la transversal. En la figura, la transversal t interseca las rectas ℓ y m. ∠1 y ∠7, y ∠2 y ∠8 son ángulos alternos externos. Si las rectas ℓ y m son paralelas, entonces estos ángulos son pares de ángulos congruentes.

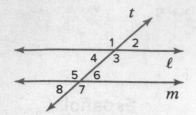

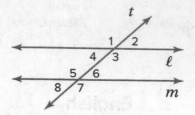

alternate interior angles Interior angles that lie on opposite sides of the transversal. In the figure below, transversal t intersects lines ℓ and m. ∠3 and ∠5, and ∠4 and ∠6 are alternate interior angles. If lines ℓ and m are parallel, then these pairs of angles are congruent.

ángulos alternos internos Ángulos internos que se encuentran en lados opuestos de la transversal. En la figura, la transversal t interseca las rectas ℓ y m. ∠3 y ∠5, y ∠4 y ∠6 son ángulos alternos internos. Si las rectas ℓ y m son paralelas, entonces estos ángulos son pares de ángulos congruentes.

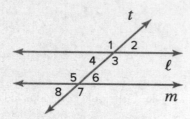

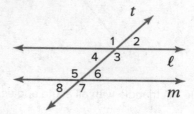

angle of rotation The degree measure of the angle through which a figure is rotated.

ángulo de rotación Medida en grados del ángulo sobre el cual se rota una figura.

arc One of two parts of a circle separated by a central angle.

arco Una de dos partes de un círculo separadas por un ángulo central.

Associative Property The way in which three numbers are grouped when they are added or multiplied does not change their sum or product.

propiedad asociativa La forma en que se agrupan tres números al sumarlos o multiplicarlos no altera su suma o producto.

Bb

base In a power, the number that is the common factor. In 10^3, the base is 10. That is, $10^3 = 10 \times 10 \times 10$.

base En una potencia, número que es el factor común. En 10^3, la base es 10. Es decir, $10^3 = 10 \times 10 \times 10$.

base One of the two parallel congruent faces of a prism.

base Una de las dos caras paralelas congruentes de un prisma.

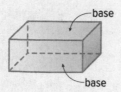

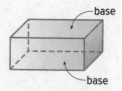

biased sample A sample drawn in such a way that one or more parts of the population are favored over others.

bivariate data Data with two variables, or pairs of numerical observations.

box plot A method of visually displaying a distribution of data values by using the median, quartiles, and extremes of the data set. A box shows the middle 50% of the data.

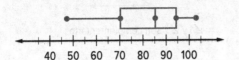

muestra sesgada Muestra en que se favorece una o más partes de una población.

datos bivariantes Datos con dos variables, o pares de observaciones numéricas.

diagrama de caja Un método de mostrar visualmente una distribución de valores usando la mediana, cuartiles y extremos del conjunto de datos. Una caja muestra el 50% del medio de los datos.

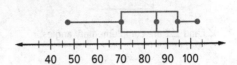

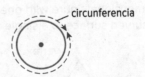

Cc

center The given point from which all points on a circle are the same distance.

center of dilation The center point from which dilations are performed.

center of rotation A fixed point around which shapes move in a circular motion to a new position.

central angle An angle that intersects a circle in two points and has its vertex at the center of the circle.

circle The set of all points in a plane that are the same distance from a given point called the center.

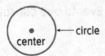

circumference The distance around a circle.

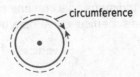

chord A segment with endpoints that are on a circle.

coefficient The numerical factor of a term that contains a variable.

common difference The difference between any two consecutive terms in an arithmetic sequence.

centro Un punto dado del cual equidistan todos los puntos de un círculo.

centro de la homotecia Punto fijo en torno al cual se realizan las homotecias.

centro de rotación Punto fijo alrededor del cual se giran las figuras en movimiento circular alrededor de un punto fijo.

ángulo central Ángulo que interseca un círculo en dos puntos y cuyo vértice es el centro del círculo.

círculo Conjunto de todos los puntos en un plano que equidistan de un punto dado llamado centro.

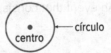

circunferencia La distancia alrededor de un círculo.

cuerda Segmento cuyos extremos están sobre un círculo.

coeficiente Factor numérico de un término que contiene una variable.

diferencia común La diferencia entre cualquier par de términos consecutivos en una sucesión aritmética.

Commutative Property   The order in which two numbers are added or multiplied does not change their sum or product.

propiedad conmutativa   La forma en que se suman o multiplican dos números no altera su suma o producto.

complementary angles   Two angles are complementary if the sum of their measures is 90°.

ángulos complementarios   Dos ángulos son complementarios si la suma de sus medidas es 90°.

∠1 and ∠2 are complementary angles.

∠1 y ∠2 son complementarios.

composite figure   A figure that is made up of two or more shapes.

figura compleja   Figura compuesta de dos o más formas.

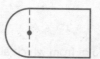

composite solid   An object made up of more than one type of solid.

sólido complejo   Cuerpo compuesto de más de un tipo de sólido.

composition of transformations   The resulting transformation when a transformation is applied to a figure and then another transformation is applied to its image.

composición de transformaciones   Transformación que resulta cuando se aplica una transformación a una figura y luego se le aplica otra transformación a su imagen.

compound event   An event that consists of two or more simple events.

evento compuesto   Evento que consta de dos o más eventos simples.

compound interest   Interest paid on the initial principal and on interest earned in the past.

interés compuesto   Interés que se paga por el capital inicial y sobre el interés ganado en el pasado.

cone   A three-dimensional figure with one circlular base connected by a curved surface to a single vertex.

cono   Una figura tridimensional con una circlular base conectada por una superficie curva para un solo vértice.

vertex

vértice

congruent   Having the same measure; if one image can be obtained by another by a sequence of rotations, reflections, or translations.

congruente   Que tienen la misma medida; si una imagen puede obtenerse de otra por una secuencia de rotaciones, reflexiones o traslaciones.

constant   A term without a variable.

constante   Término sin variables.

constant of proportionality   The constant ratio in a proportional linear relationship.

constante de proporcionalidad   La razón constante en una relación lineal proporcional.

constant of variation   A constant ratio in a direct variation.

constante de variación   Razón constante en una relación de variación directa.

constant rate of change   The rate of change between any two points in a linear relationship is the same or *constant.*

tasa constante de cambio   La tasa de cambio entre dos puntos cualesquiera en una relación lineal permanece igual o *constante.*

continuous data   Data that can take on any value. There is no space between data values for a given domain. Graphs are represented by solid lines.

datos continuos   Datos que pueden tomar cualquier valor. No hay espacio entre los valores de los datos para un dominio dado. Las gráficas se representan con rectas sólidas.

convenience sample   A sample which includes members of the population that are easily accessed.

muestra de conveniencia   Muestra que incluye miembros de una población fácilmente accesibles.

converse   The converse of a theorem is formed when the parts of the theorem are reversed. The converse of the Pythagorean Theorem can be used to test whether a triangle is a right triangle. If the sides of the triangle have lengths a, b, and c, such that $c^2 = a^2 + b^2$, then the triangle is a right triangle.

recíproco   El recíproco de un teorema se forma cuando se invierten las partes del teorema. El recíproco del teorema de Pitágoras puede usarse para averiguar si un triángulo es un triángulo rectángulo. Si las longitudes de los lados de un triángulo son a, b y c, tales que $c^2 = a^2 + b^2$, entonces el triángulo es un triángulo rectángulo.

coordinate plane   A coordinate system in which a horizontal number line and a vertical number line intersect at their zero points.

plano de coordenadas   Sistema de coordenadas en que una recta numérica horizontal y una recta numérica vertical se intersecan en sus puntos cero.

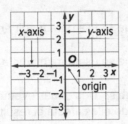

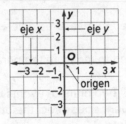

coplanar   Lines that lie in the same plane.

coplanario   Rectas que yacen en el mismo plano.

corresponding angles   Angles that are in the same position on two parallel lines in relation to a transversal.

ángulos correspondientes   Ángulos que están en la misma posición sobre dos rectas paralelas en relación con la transversal.

corresponding parts   Parts of congruent or similar figures that match.

partes correspondientes   Partes de figuras congruentes o semejantes que coinciden.

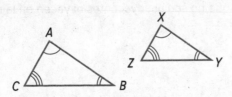

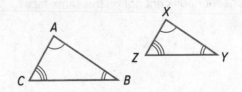

counterexample A statement or example that shows a conjecture is false.

contraejemplo Ejemplo o enunciado que demuestra que una conjetura es falsa.

cross section The intersection of a solid and a plane.

sección transversal Intersección de un sólido y un plano.

cube root One of three equal factors of a number. If $a^3 = b$, then a is the cube root of b. The cube root of 64 is 4 since $4^3 = 64$.

raíz cúbica Uno de tres factores iguales de un número. Si $a^3 = b$, entonces a es la raíz cúbica de b. La raíz cúbica de 64 es 4, dado que $4^3 = 64$.

cylinder A three-dimensional figure with two parallel congruent circular bases connected by a curved surface.

cilindro Una figura tridimensional con dos paralelas congruentes circulares bases conectados por una superficie curva.

Dd

deductive reasoning A system of reasoning that uses facts, rules, definitions, or properties to reach logical conclusions.

razonamiento deductivo Sistema de razonamiento que emplea hechos, reglas, definiciones o propiedades para obtener conclusions lógicas.

defining a variable Choosing a variable and a quantity for the variable to represent in an expression or equation.

definir una variable El elegir una variable y una cantidad que esté representada por la variable en una expresión o en una ecuación.

degree A unit used to measure angles.

grado Unidad que se usa para medir ángulos.

degree A unit used to measure temperature.

grado Unidad que se usa para medir la temperatura.

dependent events Two or more events in which the outcome of one event does affect the outcome of the other event or events.

eventos dependientes Dos o más eventos en que el resultado de uno de ellos afecta el resultado de los otros eventos.

dependent variable The variable in a relation with a value that depends on the value of the independent variable.

variable dependiente La variable en una relación cuyo valor depende del valor de la variable independiente.

derived unit A unit that is derived from a measurement system base unit, such as length, mass, or time.

unidad derivada Unidad derivada de una unidad básica de un sistema de medidas como por ejemplo, la longitud, la masa o el tiempo.

diagonal A line segment whose endpoints are vertices that are neither adjacent nor on the same face.

diagonal Segmento de recta cuyos extremos son vértices que no son ni adyacentes ni yacen en la misma cara.

diameter The distance across a circle through its center.

diameter

diámetro La distancia a través de un círculo pasando por el centro.

diámetro

dilation A transformation that enlarges or reduces a figure by a scale factor.

homotecia Transformación que produce la ampliación o reducción de una imagen por un factor de escala.

dimensional analysis The process of including units of measurement when you compute.

análisis dimensional Proceso que incorpora las unidades de medida al hacer cálculos.

direct variation A relationship between two variable quantities with a constant ratio.

variación directa Relación entre dos cantidades variables con una razón constante.

discount The amount by which a regular price is reduced.

descuento La cantidad de reducción del precio normal.

discrete data Data with space between possible data values. Graphs are represented by dots.

datos discretos Datos con espacios entre posibles valores de datos. Las gráficas están representadas por puntos.

disjoint events Events that cannot happen at the same time.

eventos disjuntos Eventos que no pueden ocurrir al mismo tiempo.

Distance Formula The distance d between two points with coordinates (x_1, y_1) and (x_2, y_2) is given by the formula

$$d = \sqrt{(x_1 - x_2)^2 + (y_1 - y_2)^2}.$$

fórmula de la distancia La distancia d entre dos puntos con coordenadas (x_1, y_1) and (x_2, y_2) viene dada por la fórmula

$$d = \sqrt{(x_1 - x_2)^2 + (y_1 - y_2)^2}.$$

distribution A way to show the arrangement of data values.

distribución Una manera de mostrar la agrupación de valores.

Distributive Property To multiply a sum by a number, multiply each addend by the number outside the parentheses.

$$5(x + 3) = 5x + 15$$

propiedad distributiva Para multiplicar una suma por un número, multiplica cada sumando por el número fuera de los paréntesis.

$$5(x + 3) = 5x + 15$$

Division Property of Equality If you divide each side of an equation by the same nonzero number, the two sides remain equal.

propiedad de división de la igualdad Si cada lado de una ecuación se divide entre el mismo número no nulo, los dos lados permanecen iguales.

domain The set of x-coordinates in a relation.

dominio Conjunto de coordenadas x en una relación.

double box plot Two box plots graphed on the same number line.

doble diagrama de puntos Dos diagramas de caja sobre la misma recta numérica.

edge The line segment where two faces of a polyhedron intersect.

equation A mathematical sentence stating that two quantities are equal.

equiangular A polygon in which all angles are congruent.

equilateral triangle A triangle with three congruent sides.

equivalent expressions Expressions that have the same value regardless of the value(s) of the variable(s).

event An outcome is a possible result.

experimental probability An estimated probability based on the relative frequency of positive outcomes occurring during an experiment.

exponent In a power, the number of times the base is used as a factor. In 10^3, the exponent is 3.

exponential function A nonlinear function in which the base is a constant and the exponent is an independent variable.

exterior angles The four outer angles formed by two lines cut by a transversal.

arista El segmento de línea donde se cruzan dos caras de un poliedro.

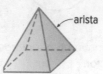

ecuación Enunciado matemático que establece que dos cantidades son iguales.

equiangular Polígono en el cual todos los ángulos son congruentes.

triángulo equilátero Triángulo con tres lados congruentes.

expresiones equivalentes Expresiones que poseen el mismo valor, sin importar los valores de la(s) variable(s).

evento Un resultado posible.

probabilidad experimental Probabilidad estimada que se basa en la frecuencia relativa de los resultados positivos que ocurren durante un experimento.

exponente En una potencia, el número de veces que la base se usa como factor. En 10^3, el exponente es 3.

función exponencial Función no lineal en la cual la base es una constante y el exponente es una variable independiente.

ángulo externo Los cuatro ángulos exteriores que se forman cuando una transversal corta dos rectas.

face A flat surface of a polyhedron.

cara Una superficie plana de un poliedro.

fair game   A game where each player has an equally likely chance of winning.

juego justo   Juego donde cada jugador tiene igual posibilidad de ganar.

five-number summary   A way of characterizing a set of data that includes the minimum, first quartile, median, third quartile, and the maximum.

resumen de los cinco números   Una manera de caracterizar un conjunto de datos que incluye el mínimo, el primer cuartil, la mediana, el tercer cuartil y el máximo.

formal proof   A two-column proof containing statements and reasons.

demonstración formal   Demonstración endos columnas contiene enunciados y razonamientos.

function   A relation in which each member of the domain (input value) is paired with exactly one member of the range (output value).

función   Relación en la cual a cada elemento del dominio (valor de entrada) le corresponde exactamente un único elemento del rango (valor de salida).

function table   A table organizing the domain, rule, and range of a function.

tabla de funciones   Tabla que organiza la regla de entrada y de salida de una función.

Fundamental Counting Principle   Uses multiplication of the number of ways each event in an experiment can occur to find the number of possible outcomes in a sample space.

principio fundamental de contar   Método que usa la multiplicación del número de maneras en que cada evento puede ocurrir en un experimento, para calcular el número de resultados posibles en un espacio muestral.

Gg

geometric sequence   A sequence in which each term after the first is found by multiplying the previous term by a constant.

sucesión geométrica   Sucesión en la cual cada término después del primero se determina multiplicando el término anterior por una constante.

Hh

half-plane   The part of the coordinate plane on one side of the boundary.

semiplano   Parte del plano de coordenadas en un lado de la frontera.

hemisphere   One of two congruent halves of a sphere.

hemisferio   Una de dos mitades congruentes de una esfera.

hypotenuse   The side opposite the right angle in a right triangle.

hipotenusa   El lado opuesto al ángulo recto de un triángulo rectángulo.

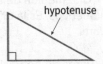

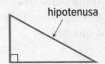

Ii

identity   An equation that is true for every value for the variable.

identidad   Ecuación que es verdad para cada valor de la variable.

image The resulting figure after a transformation.

imagen Figura que resulta después de una transformación.

independent events Two or more events in which the outcome of one event does not affect the outcome of the other event(s).

eventos independientes Dos o más eventos en los cuales el resultado de un evento no afecta el resultado de los otros eventos.

independent variable The variable in a function with a value that is subject to choice.

variable independiente Variable en una función cuyo valor está sujeto a elección.

indirect measurement A technique using properties of similar polygons to find distances or lengths that are difficult to measure directly.

medición indirecta Técnica que usa las propiedades de polígonos semejantes para calcular distancias o longitudes difíciles de medir directamente.

inductive reasoning Reasoning that uses a number of specific examples to arrive at a plausible generalization or prediction. Conclusions arrived at by inductive reasoning lack the logical certainty of those arrived at by deductive reasoning.

razonamiento inductivo Razonamiento que usa varios ejemplos especificos para lograr una generalización o una predicción plausible. Las conclusions obtenidas por razonamiento inductivo carecen de la certeza lógica de aquellas obtenidas por razonamiento deductivo.

inequality A mathematical sentence that contains <, >, ≠, ≤, or ≥.

desigualdad Enunciado matemático que contiene <, >, ≠, ≤, o ≥.

inscribed angle An angle that has its vertex on the circle. Its sides contain chords of the circle.

ángulo inscrito Ángulo cuyo vértice está en el círculo y cuyos lados contienen cuerdas del círculo.

informal proof A paragraph proof.

demonstración informal Demonstración en forma de párrafo.

interest The amount of money paid or earned for the use of money.

interés Cantidad que se cobra o se paga por el uso del dinero.

interior angle An angle inside a polygon.

ángulo interno Ángulo dentro de un polígono.

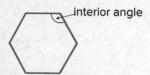

interior angle

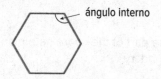

ángulo interno

interior angles The four inside angles formed by two lines cut by a transversal.

ángulo interno Los cuatro ángulos internos formados por dos rectas intersecadas por una transversal.

interquartile range A measure of variation in a set of numerical data. It is the difference between the first quartile and the third quartile.

rango intercuartílico Una medida de la variación en un conjunto de datos numéricos. Es la diferencia entre el primer y el tercer cuartil.

inverse operations Pairs of operations that undo each other. Addition and subtraction are inverse operations. Multiplication and division are inverse operations.

peraciones inversas Pares de operaciones que se anulan mutuamente. La adición y la sustracción son operaciones inversas. La multiplicación y la división son operaciones inversas.

irrational number A number that cannot be expressed as the quotient $\frac{a}{b}$, where a and b are integers and $b \neq 0$.

números irracionales Número que no se puede expresar como el cociente $\frac{a}{b}$, donde a y b son enteros y $b \neq 0$.

isosceles triangle   A triangle with at least two congruent sides.

triángulo isóceles   Triángulo con por lo menos dos lados congruentes.

lateral area   The sum of the areas of the lateral faces of a solid.

área lateral   La suma de las áreas de las caras laterales de un sólido.

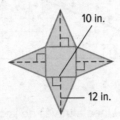

lateral area = $4\left(\frac{1}{2} \times 10 \times 12\right) = 240$ square inches

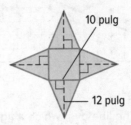

área lateral = $4\left(\frac{1}{2} \times 10 \times 12\right) = 240$ pulgadas cuadradas

lateral face   Any flat surface that is not a base.

cara lateral   Cualquier superficie plana que no es la base.

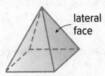

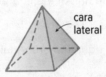

legs   The two sides of a right triangle that form the right angle.

catetos   Los dos lados de un triángulo rectángulo que forman el ángulo recto.

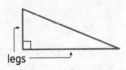

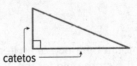

like fractions   Fractions that have the same denominators.

fracciones semejantes   Fracciones que tienen el mismo denominador.

like terms   Terms that contain the same variable(s) to the same powers.

términos semejantes   Términos que contienen la misma variable o variables elevadas a la misma potencia.

linear   To fall in a straight line.

lineal   Que cae en una línea recta.

linear equation   An equation with a graph that is a straight line.

ecuación lineal   Ecuación cuya gráfica es una recta.

linear function   A function in which the graph of the solutions forms a line.

función lineal   Función en la cual la gráfica de las soluciones forma un recta.

linear relationship A relationship that has a straight-line graph.

relación lineal Relación cuya gráfica es una recta.

line of best fit A line that is very close to most of the data points in a scatter plot.

recta de mejor ajuste Recta que más se acerca a la mayoría de puntos de los datos en un diagrama de dispersión.

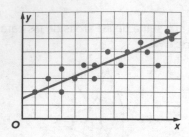

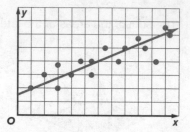

line of reflection The line over which a figure is reflected.

línea de reflexión Línea a través de la cual se refleja una figura.

line of symmetry Each half of a figure is a mirror image of the other half when a line of symmetry is drawn.

eje de simetría Recta que divide una figura en dos mitades especulares.

line symmetry A figure has line symmetry if a line can be drawn so that one half of the figure is a mirror image of the other half.

simetría lineal Una figura tiene simetría lineal si se puede trazar una recta de manera que una mitad de la figura sea una imagen especular de la otra mitad.

literal equation An equation or formula that has more than one variable.

ecuación literal Ecuación o fórmula con más de una variable.

Mm

markup The amount the price of an item is increased above the price the store paid for the item.

margen de utilidad Cantidad de aumento en el precio de un artículo por encima del precio que paga la tienda por dicho artículo.

mean The sum of the data divided by the number of items in the set.

media La suma de datos dividida entre el número total de artículos.

mean absolute deviation The average of the absolute values of differences between the mean and each value in a data set.

desviación media absoluta El promedio de los valores absolutos de diferencias entre el medio y cada valor de un conjunto de datos.

measures of center Numbers that are used to describe the center of a set of data. These measures include the mean, median, and mode.

medidas del centro Números que describen el centro de un conjunto de datos. Estas medidas incluyen la media, la mediana y la moda.

measures of variation Numbers used to describe the distribution or spread of a set of data.

medidas de variación Números que se usan para describir la distribución o separación de un conjunto de datos.

median   A measure of center in a set of numerical data. The median of a list of values is the value appearing at the center of a sorted version of the list—or the mean of the two central values, if the list contains an even number of values.

mode   The number(s) or item(s) that appear most often in a set of data.

monomial   A number, a variable, or a product of a number and one or more variables.

Multiplication Property of Equality   If you multiply each side of an equation by the same number, the two sides remain equal.

multiplicative inverses   Two numbers with a product of 1. The multiplicative inverse of $\frac{2}{3}$ is $\frac{3}{2}$.

mediana   Una medida del centro en un conjunto de datos numéricos. La mediana de una lista de valores es el valor que aparece en el centro de una versión ordenada de la lista, o la media de los dos valores centrales si la lista contiene un número par de valores.

moda   El número(s) o artículo(s) que aparece con más frecuencia en un conjunto de datos.

monomio   Un número, una variable o el producto de un número por una o más variables.

propiedad de multiplicación de la igualdad   Si cada lado de una ecuación se multiplica por el mismo número, los lados permanecen iguales.

inversos multiplicativo   Dos números cuyo producto es 1. El inverso multiplicativo de $\frac{2}{3}$ es $\frac{3}{2}$.

net   A two-dimensional pattern of a three-dimensional figure.

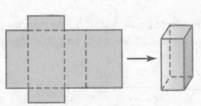

red   Patrón bidimensional de una figura tridimensional.

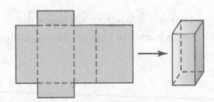

nonlinear function   A function whose rate of change is not constant. The graph of a nonlinear function is not a straight line.

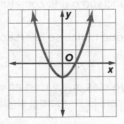

función no lineal   Función cuya tasa de cambio no es constante. La gráfica de una función no lineal no es una recta.

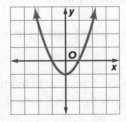

null set   The empty set.

conjunto nulo   El conjunto vacío.

obtuse angle   An angle whose measure is between 90° and 180°.

obtuse triangle   A triangle with one obtuse angle.

ángulo obtuso   Ángulo cuya medida está entre 90° y 180°.

triángulo obtusángulo   Triángulo con un ángulo obtuso.

ordered pair A pair of numbers used to locate a point in the coordinate plane. The ordered pair is written in this form: (*x*-coordinate, *y*-coordinate).

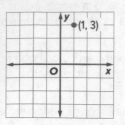

par ordenado Par de números que se utiliza para ubicar un punto en un plano de coordenadas. Se escribe de la siguiente forma: (coordenada *x*, coordenada *y*).

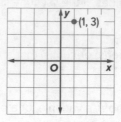

origin The point of intersection of the *x*-axis and *y*-axis in a coordinate plane.

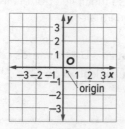

origen Punto en que el eje *x* y el eje *y* se intersecan en un plano de coordenadas.

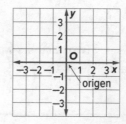

outcome One possible result of a probability event. For example, 4 is an outcome when a number cube is rolled.

resultado Una consecuencia posible de un evento de probabilidad. Por ejemplo, 4 es un resultado posible al lanzar un cubo numérico.

outlier Data that are more than 1.5 times the interquartile range from the first or third quartiles.

valor atípico Datos que distan de los cuartiles respectivos más de 1.5 veces la amplitud intercuartílica.

Pp

paragraph proof A paragraph that explains why a statement or conjecture is true.

prueba por párrafo Párrafo que explica por qué es verdadero un enunciado o una conjetura.

parallel Lines that never intersect no matter how far they extend.

paralelo Rectas que nunca se intersecan sea cual sea su extensión.

parallel lines Lines in the same plane that never intersect or cross. The symbol ∥ means parallel.

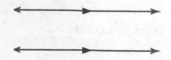

rectas paralelas Rectas que yacen en un mismo plano y que no se intersecan. El símbolo ∥ significa paralela a.

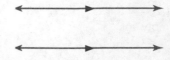

parallelogram A quadrilateral with both pairs of opposite sides parallel and congruent.

paralelogramo Cuadrilátero con ambos pares de lados opuestos, paralelos y congruentes.

percent equation   An equivalent form of a percent proportion in which the percent is written as a decimal.

$$part = percent \cdot whole$$

percent of change   A ratio that compares the change in quantity to the original amount.

$$percent\ of\ change = \frac{amount\ of\ change}{original\ amount}$$

percent of decrease   When the percent of change is negative.

percent of increase   When the percent of change is positive.

percent proportion   Compares part of a quantity to the whole quantity using a percent.

$$\frac{part}{whole} = \frac{percent}{100}$$

perfect cube   A rational number whose cube root is a whole number. 27 is a perfect cube because its cube root is 3.

perfect square   A rational number whose square root is a whole number. 25 is a perfect square because its square root is 5.

permutation   An arrangement or listing in which order is important.

perpendicular lines   Two lines that intersect to form right angles.

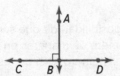

pi   The ratio of the circumference of a circle to its diameter. The Greek letter π represents this number. The value of pi is always 3.1415926... .

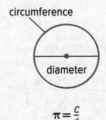

$$\pi = \frac{C}{d}$$

point-slope form   An equation of the form $y - y_1 = m(x - x_1)$, where m is the slope and (x_1, y_1) is a given point on a nonvertical line.

ecuación porcentual   Forma equivalente de proporción porcentual en la cual el por ciento se escribe como un decimal.

$$parte = por\ ciento \cdot entero$$

porcentaje de cambio   Razón que compara el cambio en una cantidad a la cantidad original.

$$procentaje\ de\ cambio = \frac{cantidad\ de\ cambio}{cantidad\ original}$$

porcentaje de disminución   Cuando el porcentaje de cambio es negativo.

porcentaje de aumento   Cuando el porcentaje de cambio es positivo.

proporción porcentual   Compara parte de una cantidad con la cantidad total mediante un por ciento.

$$\frac{parte}{entero} = \frac{por\ ciento}{100}$$

cubo perfecto   Número racional cuya raíz cúbica es un número entero. 27 es un cubo perfecto porque su raíz cúbica es 3.

cuadrados perfectos   Número racional cuya raíz cuadrada es un número entero. 25 es un cuadrado perfecto porque su raíz cuadrada es 5.

permutación   Arreglo o lista donde el orden es importante.

rectas perpendiculares   Dos rectas que se intersecan formando ángulos rectos.

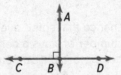

pi   Razón de la circunferencia de un círculo al diámetro del mismo. La letra griega π representa este número. El valor de pi es siempre 3.1415926... .

$$\pi = \frac{C}{d}$$

forma punto-pendiente   Ecuación de la forma $y - y_1 = m(x - x_1)$ donde m es la pendiente y $(x_1 - y_1)$ es un punto dado de una recta no vertical.

polygon   A simple, closed figure formed by three or more line segments.

poligono   Figura simple y cerrada formada por tres o más segmentos de recta.

polyhedron   A three-dimensional figure with faces that are polygons.

poliedro   Una figura tridimensional con caras que son polígonos.

power   A product of repeated factors using an exponent and a base. The power 7^3 is read *seven to the third power,* or *seven cubed.*

potencia   Producto de factores repetidos con un exponente y una base. La potencia 7^3 se lee *siete a la tercera potencia* o *siete al cubo.*

precision   The ability of a measurement to be consistently reproduced.

precisión   Capacidad de una medida a ser reproducida consistentemente.

preimage   The original figure before a transformation.

preimagen   Figura original antes de una transformación.

principal   The amount of money invested or borrowed.

capital   Cantidad de dinero que se invierte o que se toma prestada.

prism   A polyhedron with two parallel congruent faces called bases.

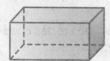

prisma   Poliedro con dos caras congruentes y paralelas llamadas bases.

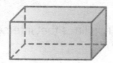

probability   The chance that some event will happen. It is the ratio of the number of ways a certain event can occur to the number of possible outcomes.

probabilidad   La posibilidad de que suceda un evento. Es la razón del número de maneras en que puede ocurrir un evento al número total de resultados posibles.

proof   A logical argument in which each statement that is made is supported by a statement that is accepted as true.

prueba   Argumento lógico en el cual cada enunciado hecho se respalda con un enunciado que se acepta como verdadero.

property   A statement that is true for any numbers.

propiedad   Enunciado que se cumple para cualquier número.

pyramid   A polyhedron with one base that is a polygon and three or more triangular faces that meet at a common vertex.

pirámide   Un poliedro con una base que es un polígono y tres o más caras triangulares que se encuentran en un vértice común.

Pythagorean Theorem In a right triangle, the square of the length of the hypotenuse c is equal to the sum of the squares of the lengths of the legs a and b.
$a^2 + b^2 = c^2$

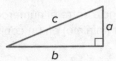

Teorema de Pitágoras En un triángulo rectángulo, el cuadrado de la longitud de la hipotenusa es igual a la suma de los cuadrados de las longitudes de los catetos.
$a^2 + b^2 = c^2$

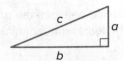

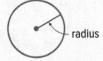

Qq

quadrants The four sections of the coordinate plane.

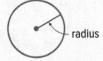

cuadrantes Las cuatro secciones del plano de coordenadas.

quadratic function A function in which the greatest power of the variable is 2.

función cuadrática Función en la cual la potencia mayor de la variable es 2.

quadrilateral A closed figure with four sides and four angles.

cuadrilátero Figura cerrada con cuatro lados y cuatro ángulos.

qualitative graph A graph used to represent situations that do not necessarily have numerical values.

gráfica cualitativa Gráfica que se usa para representar situaciones que no tienen valores numéricos necesariamente.

quantitative data Data that can be given a numerical value.

datos cualitativos Datos que se pueden dar un valor numérico.

quartiles Values that divide a set of data into four equal parts.

cuartiles Valores que dividen un conjunto de datos en cuatro partes iguales.

Rr

radical sign The symbol used to indicate a positive square root, $\sqrt{}$.

signo radical Símbolo que se usa para indicar una raíz cuadrada no positiva, $\sqrt{}$.

radius The distance from the center of a circle to any point on the circle.

radio Distancia desde el centro de un círculo hasta cualquier punto del mismo.

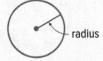

radius

radio

random   Outcomes occur at random if each outcome is equally likely to occur.

azar   Los resultados ocurren al azar si todos los resultados son equiprobables.

range   The set of y-coordinates in a relation.

rango   Conjunto de coordenadas y en una relación.

range   The difference between the greatest number (maximum) and the least number (minimum) in a set of data.

rango   La diferencia entre el número mayor (máximo) y el número menor (mínimo) en un conjunto de datos.

rational number   Numbers that can be written as the ratio of two integers in which the denominator is not zero. All integers, fractions, mixed numbers, and percents are rational numbers.

número racional   Números que pueden escribirse como la razón de dos enteros en los que el denominador no es cero. Todos los enteros, fracciones, números mixtos y porcentajes son números racionales.

real numbers   The set of rational numbers together with the set of irrational numbers.

número real   El conjunto de números racionales junto con el conjunto de números irracionales.

reciprocals   The multiplicative inverse of a number. The product of reciprocals is 1.

recíproco   El inverso multiplicativo de un número. El producto de recíprocos es 1.

reflection   A transformation where a figure is flipped over a line. Also called a flip.

reflexión   Transformación en la cual una figura se voltea sobre una recta. También se conoce como simetría de espejo.

regular polygon   A polygon that is equilateral and equiangular.

polígono regular   Polígono equilátero y equiangular.

regular pyramid   A pyramid whose base is a regular polygon.

pirámide regular   Pirámide cuya base es un polígono regular.

relation   Any set of ordered pairs.

relación   Cualquier conjunto de pares ordenados.

relative frequency   The ratio of the number of experimental successes to the total number of experimental attempts.

frecuencia relativa   Razón del número de éxitos experimentales al número total de intentos experimentales.

remote interior angles   The angles of a triangle that are not adjacent to a given exterior angle.

ángulos internos no adyacentes   Ángulos de un triángulo que no son adya centes a un ángulo exterior dado.

repeating decimal   Decimal form of a rational number.

decimal periódico   Forma decimal de un número racional.

rhombus   A parallelogram with four congruent sides.

rombo   Paralelogramo con cuatro lados congruentes.

right angle   An angle whose measure is exactly 90°.

right triangle   A triangle with one right angle.

rise   The vertical change between any two points on a line.

rotation   A transformation in which a figure is turned about a fixed point.

rotational symmetry   A type of symmetry a figure has if it can be rotated less than 360° about its center and still look like the original.

run   The horizontal change between any two points on a line.

ángulo recto   Ángulo que mide exactamente 90°.

triángulo rectángulo   Triángulo con un ángulo recto.

elevación   El cambio vertical entre cualquier par de puntos en una recta.

rotación   Transformación en la cual una figura se gira alrededor de un punto fijo.

simetría rotacional   Tipo de simetría que tiene una figura si se puede girar menos que 360° en torno al centro y aún sigue viéndose como la figura original.

carrera   El cambio horizontal entre cualquier par de puntos en una recta.

Ss

sales tax   An additional amount of money charged on certain goods and services.

sample   A randomly-selected group chosen for the purpose of collecting data.

sample space   The set of all possible outcomes of a probability experiment.

scale factor   The ratio of the lengths of two corresponding sides of two similar polygons.

impuesto sobre las ventas   Cantidad de dinero adicional que se cobra por ciertos artículos y servicios.

muestra   Subconjunto de una población que se usa con el propósito de recoger datos.

espacio muestral   Conjunto de todos los resultados posibles de un experimento de probabilidad.

factor de escala   La razón de las longitudes de dos lados correspondientes de dos polígonos semejantes.

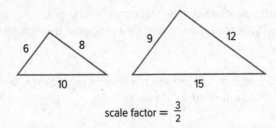

scale factor = $\frac{3}{2}$

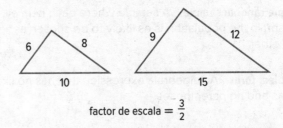

factor de escala = $\frac{3}{2}$

scalene triangle   A triangle with no congruent sides.

triángulo escaleno   Triángulo sin lados congruentes.

scatter plot   A graph that shows the relationship between a data set with two variables graphed as ordered pairs on a coordinate plane.

diagrama de dispersión   Gráfica que muestra la relación entre un conjunto de datos con dos variables graficadas como pares ordenados en un plano de coordenadas.

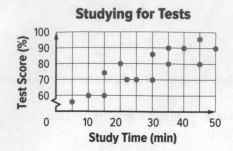

Studying for Tests

Tiempo de estudio para prueb

scientific notation   A compact way of writing numbers with absolute values that are very large or very small. In scientific notation, 5,500 is 5.5×10^3.

notación científica   Manera abreviada de escribir números con valores absolutos que son muy grandes o muy pequeños. En notación científica, 5,500 es 5.5×10^3.

selling price   The amount the customer pays for an item.

precio de venta   Cantidad de dinero que paga un consumidor por un artículo.

semicircle   An arc measuring 180°.

semicírculo   Arco que mide 180°.

sequence   An ordered list of numbers, such as 0, 1, 2, 3 or 2, 4, 6, 8.

sucesión   Lista ordenada de números, tales como 0, 1, 2, 3 o 2, 4, 6, 8.

similar   If one image can be obtained from another by a sequence of transformations and dilations.

similar   Si una imagen puede obtenerse de otra mediante una secuencia de transformaciones y dilataciones.

similar polygons   Polygons that have the same shape.

polígonos semejantes   Polígonos con la misma forma.

similar solids   Solids that have exactly the same shape, but not necessarily the same size.

sólidos semejantes   Sólidos que tienen exactamente la misma forma, pero no necesariamente el mismo tamaño.

simple interest   Interest paid only on the initial principal of a savings account or loan.

interés simple   Interés que se paga sólo sobre el capital inicial de una cuenta de ahorros o préstamo.

simple random sample   A sample where each item or person in the population is as likely to be chosen as any other.

muestra aleatoria simple   Muestra de una población que tiene la misma probabilidad de escogerse que cualquier otra.

simplest form   An algebraic expression that has no like terms and no parentheses.

forma reducida   Expresión algebraica que carece de términos semejantes y de paréntesis.

simplify   To perform all possible operations in an expression.

simplificar   Realizar todas las operaciones posibles en una expresión.

simulation   An experiment that is designed to model the action in a given situation.

simulacro   Un experimento diseñado para modelar la acción en una situación dada.

slant height   The altitude or height of each lateral face of a pyramid.

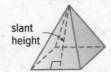

slant height

altura oblicua   La longitud de la altura de cada cara lateral de una pirámide.

altura oblicua

slope   The rate of change between any two points on a line. The ratio of the rise, or vertical change, to the run, or horizontal change.

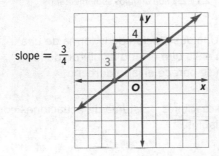

slope $= \frac{3}{4}$

pendiente   Razón de cambio entre cualquier par de puntos en una recta. La razón de la altura, o cambio vertical, a la carrera, o cambio horizontal.

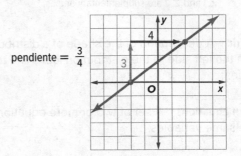

pendiente $= \frac{3}{4}$

slope-intercept form   An equation written in the form $y = mx + b$, where m is the slope and b is the y-intercept.

forma pendiente intersección   Ecuación de la forma $y = mx + b$, donde m es la pendiente y b es la intersección y.

solid   A three-dimensional figure formed by intersecting planes.

sólido   Figura tridimensional formada por planos que se intersecan.

sphere   The set of all points in space that are a given distance from a given point called the center.

esfera   Conjunto de todos los puntos en el espacio que están a una distancia dada de un punto dado llamado centro.

square root   One of the two equal factors of a number. If $a^2 = b$, then a is the square root of b. A square root of 144 is 12 since $12^2 = 144$.

raíz cuadrada   Uno de dos factores iguales de un número. Si $a^2 = b$, la a es la raíz cuadrada de b. Una raíz cuadrada de 144 es 12 porque $12^2 = 144$.

standard deviation   A measure of variation that describes how the data deviates from the mean of the data.

desviación estándar   Una medida de variación que describe cómo los datos se desvía de la media de los datos.

standard form   An equation written in the form $Ax + By = C$.

forma estándar   Ecuación escrita en la forma $Ax + By = C$.

straight angle   An angle whose measure is exactly 180°.

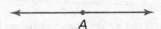

A

ángulo llano   Ángulo que mide exactamente 180°.

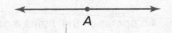

A

substitution   An algebraic model that can be used to find the exact solution of a system of equations.

sustitución   Modelo algebraico que se puede usar para calcular la solución exacta de un sistema de ecuaciones.

Subtraction Property of Equality   If you subtract the same number from each side of an equation, the two sides remain equal.

propiedad de sustracción de la igualdad   Si sustraes el mismo número de ambos lados de una ecuación, los dos lados permanecen iguales.

supplementary angles   Two angles are supplementary if the sum of their measures is 180°.

ángulos suplementarios   Dos ángulos son suplementarios si la suma de sus medidas es 180°.

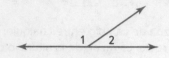

∠1 and ∠2 are supplementary angles.

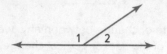

∠1 y ∠2 son ángulos suplementarios.

symmetric   A description of the shape of a distribution in which the left side of the distribution looks like the right side.

simétrico   Una descripción de la forma de una distribución en la que el lado izquierdo de la distribución se parece el lado derecho.

system of equations   A set of two or more equations with the same variables.

sistema de ecuaciones   Sistema de ecuaciones con las mismas variables.

term   A number, a variable, or a product of numbers and variables.

término   Un número, una variable o un producto de números y variables.

term   Each part of an algebraic expression separated by an addition or subtraction sign.

término   Cada parte de un expresión algebraica separada por un signo adición o un signo sustracción.

terminating decimal   A repeating decimal where the repeating digit is zero.

decimal finito   Un decimal periódico donde el dígito que se repite es cero.

theorem   A statement or conjecture that can be proven.

teorema   Un enunciado o conjetura que puede probarse.

theoretical probability   Probability based on known characteristics or facts.

probabilidad teórica   Probabilidad que se basa en características o hechos conocidos.

third quartile   For a data set with median M, the third quartile is the median of the data values greater than M.

tercer cuartil   Para un conjunto de datos con la mediana M, el tercer cuartil es la mediana de los valores mayores que M.

three-dimensional figure   A figure with length, width, and height.

figura tridimensional   Figura que tiene largo, ancho y alto.

total surface area   The sum of the areas of the surfaces of a solid.

área de superficie total   La suma del área de las superficies de un sólido.

transformation   An operation that maps a geometric figure, preimage, onto a new figure, image.

transformación   Operación que convierte una figura geométrica, la pre-imagen, en una figura nueva, la imagen.

translation A transformation that slides a figure from one position to another without turning.

transversal A line that intersects two or more other lines.

trapezoid A quadrilateral with exactly one pair of parallel sides.

tree diagram A diagram used to show the total number of possible outcomes in a probability experiment.

triangle A figure formed by three line segments that intersect only at their endpoints.

two-column proof A formal proof that contains statements and reasons organized in two columns. Each step is called a statement, and the properties that justify each step are called reasons.

two-step equation An equation that contains two operations.

two-step inequality An inequality that contains two operations.

two-way table A table that shows data that pertain to two different categories.

traslación Transformación en la cual una figura se desliza de una posición a otra sin hacerla girar.

transversal Recta que interseca dos o más rectas.

trapecio Cuadrilátero con exactamente un par de lados paralelos.

diagrama de árbol Diagrama que se usa para mostrar el número total de resultados posibles en un experimento de probabilidad.

triángulo Figura formada por tres segmentos de recta que se intersecan sólo en sus extremos.

demostración de dos columnas Demonstración formal que contiene enunciados y razones organizadas en dos columnas. Cada paso se llama enunciado y las propiedades que lo justifican son las razones.

ecuación de dos pasos Ecuación que contiene dos operaciones.

desigualdad de dos pasos Desigualdad que contiene dos operaciones.

tabla de doble entrada Una tabla que muestra datos que pertenecen a dos categorías diferentes.

Uu

unbiased sample A sample that is selected so that it is representative of the entire population.

unit rate/ratio A rate or ratio with a denominator of 1.

univariate data Data with one variable.

unlike fractions Fractions whose denominators are different.

muestra no sesgada Muestra que se selecciona de modo que sea representativa de la población entera.

tasa/razón unitaria Una tasa o razón con un denominador de 1.

datos univariante Datos con una variable.

fracciones con distinto denominador Fracciones cuyos denominadores son diferentes.

Vv

variable A symbol, usually a letter, used to represent a number in mathematical expressions or sentences.

variable Un símbolo, por lo general, una letra, que se usa para representar números en expresiones o enunciados matemáticos.

vertex   The point where the sides of an angle meet.

vértice   Punto donde se encuentran los lados.

vertex   The point where three or more faces of a polyhedron intersect.

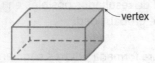

vértice   El punto donde tres o más caras de un poliedro se cruzan.

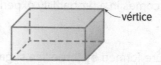

vertex   The point at the tip of a cone.

vértice   El punto en la punta de un cono.

vertical angles   Opposite angles formed by the intersection of two lines. Vertical angles are congruent. In the figure, the vertical angles are ∠1 and ∠3, and ∠2 and ∠4.

ángulos opuestos por el vértice   Ángulos congruentes que se forman de la intersección de dos rectas. En la figura, los ángulos opuestos por el vértice son ∠1 y ∠3, y ∠2 y ∠4.

volume   The measure of the space occupied by a solid. Standard measures are cubic units such as in³ or ft³.

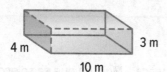

$V = 10 \times 4 \times 3 = 120$ cubic meters

volumen   Medida del espacio que ocupa un sólido. Las medidas estándares son las unidades cúbicas, como pulg³ o pies³.

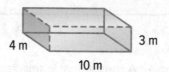

$V = 10 \times 4 \times 3 = 120$ metros cúbicos

voluntary response sample   A sample which involves only those who want to participate in the sampling.

muestra de respuesta voluntaria   Muestra que involucra sólo aquellos que quieren participar en el muestreo.

x-axis   The horizontal number line that helps to form the coordinate plane.

eje x   La recta numérica horizontal que ayuda a formar el plano de coordenadas.

x-coordinate   The first number of an ordered pair.

x-intercept   The *x*-coordinate of the point where the line crosses the *x*-axis.

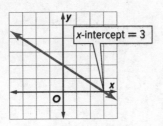

coordenada *x*   El primer número de un par ordenado.

intersección *x*   La coordenada *x* del punto donde cruza la gráfica el eje *x*.

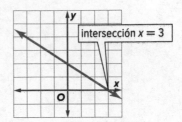

y-axis   The vertical number line that helps to form the coordinate plane.

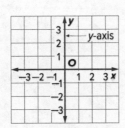

eje *y*   La recta numérica vertical que ayuda a formar el plano de coordenadas.

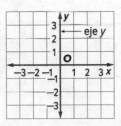

y-coordinate   The second number of an ordered pair.

y-intercept   The *y*-coordinate of the point where the line crosses the *y*-axis.

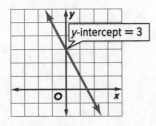

coordenada *y*   El segundo número de un par ordenado.

intersección *y*   La coordenada *y* del punto donde cruza la gráfica el eje *y*.

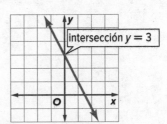

Selected Answers

Chapter 1 Real Numbers

Page 6   Chapter 1   Are You Ready?
1. 256   **3.** $2,048   **5.** $2 \times 2 \times 2 \times 3$ **7.** $2 \times 2 \times 5 \times 5$
9. $-1 \times 2 \times 3 \times 7$

Pages 11–12   Lesson 1-1   Independent Practice
1. 0.4   **3.** 0.825   **5** $-0.\overline{54}$ **7a.** $0.0\overline{6}$ **7b.** $0.1\overline{6}$ **7c.** 0.333
7d. 0.417   **9** $-7\frac{8}{25}$ **11.** $-\frac{5}{11}$ **13.** $5\frac{11}{20}$ **15.** $1\frac{1}{16}$ in.;
1.0625 in.   **17.** Sample answer: When dividing, there are two
possibilities for the remainder. If the remainder is 0, the
decimal terminates. If the remainder is not 0, then the decimal
begins to repeat at the point where the remainder repeats or
equals the original dividend.   **19.** Sample answer: 0.5, 0.555...;
0.5 < 0.555...

Pages 13–14   Lesson 1-1   Extra Practice
21. $7\frac{5}{33}$ **23.** 5.3125   **25.** $-1\frac{11}{20}$ **27.** $-\frac{1}{11}$ **29.** $2\frac{2}{5}$ in.
31. 0.45   **33a.** True   **33b.** False   **33c.** True   **33d.** False
35. >   **37.** =

Pages 19–20   Lesson 1-2   Independent Practice
1. $(-5)^4$ **3.** m^5 **5.** $\frac{1}{81}$ **7** 8,000,000,000 or 8 billion
9 -311 **11.** 16   **13a.** 10^2 **13b.** 10^6 **13c.** 10^9
13d. 10^{15} **15.** Sample answer: As the exponent decreases by
1, the simplified answer is divided by 3; $\frac{1}{2}$

Pages 21–22   Lesson 1-2   Extra Practice
17. $3^3 \cdot p^3$ **19.** $\left(-\frac{5}{6}\right)^3$ **21.** $4^2 \cdot b^4$ **23.** 224   **25.** =

27a.

Side Length (in.)	Perimeter (in.)	Area (in²)
1	4	1
2	8	4
3	12	9
4	16	16
5	20	25
6	24	36
7	28	49
8	32	64
9	36	81
10	40	100

27b.

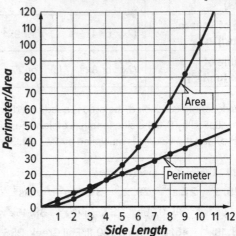

Perimeter and Area of a Square

27c. Sample answer: The graph representing perimeter of a
square is linear because each side length is multiplied by 4.
The graph representing area of a square is nonlinear because
each side length is squared and does not increase at a
constant rate.   **29.** 6^3 in³   **31.** -31 **33.** 1

Pages 27–28   Lesson 1-3   Independent Practice
1. $(-6)^7$ or $-279,936$ **3.** $-35a^6b^5c^5$ **5** $2t^3$
7. 3^3x^2 or $27x^2$ **9.** 6^5 or 7,776   **11** 10^{14} instructions
13a. 10^9 times greater   **13b.** 10^6 or one million   **15.** 9
17. 6   **19.** 7   **21.** equal; Sample answer: Using the quotient of
powers, $\frac{3^{100}}{3^{99}} = 3^{100-99}$ or 3^1, which is 3.   **23.** false; Sample
answer: If $a = 3$, then $(-3)^2 = 9$, but $-3^2 = -9$.

Pages 29–30   Lesson 1-3   Extra Practice
25. h^1 or h **27.** $-8w^{11}$ **29.** 2^8 or 256   **31.** $5^2 \cdot 7^1 \cdot 10$ or
1750   **33a.** $2r$ **33b.** $\frac{\pi}{4}$

33c.

Radius (units)	2	3	4	$2r$
Area of Circle (units²)	$\pi(2)^2$ or 4π	9π	16π	$4\pi r^2$
Length of 1 Side of the Square	4	6	8	$4r$
Area of Square (units²)	4^2 or 16	36	64	$16r^2$
Ratio (Area of circle / Area of square)	$\frac{\pi}{4}$	$\frac{\pi}{4}$	$\frac{\pi}{4}$	$\frac{\pi}{4}$

33d. The ratio is $\frac{\pi}{4}$.   **35.** Statement 1: Idaho; Statement 2:
Alabama; Statement 3: Illinois, Wyoming   **37.** 60   **39.** 3
41. -5

Pages 35–36 Lesson 1-4 Independent Practice

1. 4^6 **3.** d^{42} **5.** 3^8 **7.** $625j^{24}$ **9.** $216a^6b^{18}$
11. $-243w^{15}z^{40}$ 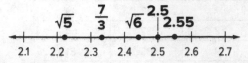 **13** $27c^{18}d^6$ in^3
15. $729x^{12}y^{18}$ **17** $-2{,}048v^{29}$

19a.

Side Length (units)	x	$2x$	$3x$
Area of Square (units^2)	x^2	$(2x)^2$ or $4x^2$	$(3x)^2$ or $9x^2$
Volume of Cube (units^3)	x^3	$(2x)^3$ or $8x^3$	$(3x)^3$ or $27x^3$

19b. If the side length is doubled, the area is quadrupled and the volume is multiplied by 8. If the side length is tripled, the area is multiplied by 9 and the volume is multiplied by 27. **21.** 3

Pages 37–38 Lesson 1-4 Extra Practice

23. 2^{14} **25.** 3^8 **27.** z^{55} **29.** 2^{18} **31.** $64g^6h^2$ units^2
33. $125r^6s^9$ units^3 **35.** $0.25k^{10}$ **37.** $\frac{1}{16}w^{10}z^6$ **39.** $16x^4$ yd^2
41. 6^{11} **43.** $18x^{14}$ **45.** Bridalveil: 620 ft; Fall Creek: 256 ft; Shoshone: 212 ft

Page 41 Problem-Solving Investigation The Four-Step Plan

Case 3. 18 tour guides **Case 5.** $2n + 1$; 201 toothpicks

Pages 47–48 Lesson 1-5 Independent Practice

1. $\frac{1}{7^{10}}$ **3.** $\frac{1}{g^7}$ **5.** 12^{-4} **7.** 5^{-3} **9.** $10^{-1}, 10^{-2}, 10^{-3}, 10^{-6}$
11. $\frac{1}{128}$ **13** y^3 **15.** 81 **17.** y^4 **19** 10^5 or 100,000 times
21. $11^{-3}, 11^0, 11^2$; Sample answer: The exponents in order from least to greatest are $-3, 0, 2$. **23.** Sample answer: $\left(\frac{1}{2}\right)^{-1} = 2, \left(\frac{34}{43}\right)^{-1} = \frac{43}{34}, \left(\frac{56}{65}\right)^{-1} = \left(\frac{65}{56}\right)$; When you raise a fraction to the -1 power, it is the same as finding the reciprocal of the fraction.

Pages 49–50 Lesson 1-5 Extra Practice

25. $\frac{1}{3^5}$ **27.** $\frac{1}{6^8}$ **29.** $\frac{1}{s^9}$ **31.** z^{-1} or $\frac{1}{z}$ **33.** b^{-12} or $\frac{1}{b^{12}}$
35. $\frac{1}{16}$ **37.** $\frac{1}{10{,}000}$ **39.** 12 **41.** -11 **43.** micrometer: 10^{-6}; millimeter: 10^{-3}; nanometer: 10^{-9}; picometer: 10^{-12}
45. 1,000 **47.** 100,000 **49.** 100 **51.** 100 **53.** 10

Pages 55–56 Lesson 1-6 Independent Practice

1. 3,160 **3.** 0.0000252 **5.** 7.2×10^{-3} **7** Arctic, Southern, Indian, Atlantic, Pacific **9.** 17.32 millimeters; the number is small so choosing a smaller unit of measure is more meaningful. **11** < **13.** 1.2×10^6; 1.2×10^5 is only 120,000, but 1.2×10^6 is just over one million. **15.** Sample answer: $3.01 \times 10^2, 5.01 \times 10^2$; $3.01 \times 10^2 < 5.01 \times 10^2$

Pages 57–58 Lesson 1-6 Extra Practice

17. 7.07×10^{-6} **19.** 0.0078 **21.** 6.7×10^3 **23.** 3.7×10^{-2}
25. 2.2×10^3, 310,000, 3.1×10^7, 216,000,000 **27a.** No
27b. Yes **27c.** Yes **27d.** No **29.** 10.232 **31.** 677.6
33. $50x^7$

Pages 63–64 Lesson 1-7 Independent Practice

1. 8.97×10^8 **3.** 8.19×10^{-2} **5** 2.375×10^{11}
7. 8,000 times **9** 9.83×10^8 **11.** 8.70366×10^4
13. $\frac{6.63 \times 10^{-6}}{5.1 \times 10^{-2}} = \left(\frac{6.63}{5.1}\right)\left(\frac{10^{-6}}{10^{-2}}\right)$
$= 1.3 \times 10^{-6 - (-2)}$
$= 1.3 \times 10^{-4}$
15a. 1×10^{100} **15b.** 10^{109} times **15c.** about 4×10^{89} adults

Pages 65–66 Lesson 1-7 Extra Practice

17. 4.44×10^1 **19.** 4×10^2 **21.** 1.334864×10^{10}
23. $13\frac{5}{9}$ h **25.** fish; mammals; 23,500 **27a.** 43.56 in^2
27b. 287.496 in^3

Pages 75–76 Lesson 1-8 Independent Practice

1. 4 **3.** no real solution **5** -1.6 **7.** ± 9 **9.** ± 0.13
11. -0.5 **13** 13 students **15.** 44 in. **17.** 24 m
19. $\frac{25}{81}$ **21.** x **23.** Sample answer: There are no two equal numbers that have a product of -4, but $(-2)(-2)(-2) = -8$.

Pages 77–78 Lesson 1-8 Extra Practice

25. -9 **27.** $-\frac{4}{5}$ **29.** -6 **31.** -10 **33.** ± 10 **35.** ± 1.1
37. 1.1 **39.** 25 **41.** 110.25 **43.** 156,816 ft^2, 174,724 ft^2
45. 2,197 **47.** 3,375 **49.** 55 **51.** 20 **53.** $64r^9s^3$ units^3

Pages 85–86 Lesson 1-9 Independent Practice

1. 5 **3.** 4 **5.** 3 **7** 10 **9.** Sample answer: 54 ft and 57 ft; 55.5 ft and 55.8 feet; 55.71 ft and 55.74 feet; 56 feet
11 about 2.75 seconds **13.** $\sqrt[3]{105}, 5, \sqrt{38}, 7$ **15.** 10; Since 94 is less than 100, $\sqrt{94}$ is less than 10. **17.** She incorrectly estimated. She found half of 200, not the square root. Since $196 < 200 < 225$, the square root of 200 is between 14 and 15. Since 200 is closer to 196, the square root of 200 is about 14. **19.** sometimes; Sample answer: $\sqrt{9}$ is greater than $\sqrt[3]{18}$, but $\sqrt{4}$ is less than $\sqrt[3]{18}$.

Pages 87–88 Lesson 1-9 Extra Practice

21. 6 **23.** 5 **25.** 8 **27.** 10 or -10 **29.** 6 in. **31.** 70 feet on each side **33.** 35 mph **35.** $\frac{17}{10}$ **37.** $\frac{49}{50}$

Pages 93–94 Lesson 1-10 Independent Practice

1. rational **3.** rational **5.** < **7** <
9 $\sqrt{5}, \frac{7}{3}, \sqrt{6}, 2.5, 2.55$

A number line marked with $\sqrt{5}$, $\frac{7}{3}$, $\sqrt{6}$, 2.5, 2.55 from 2.1 to 2.7.

11. about 1.9 m^2 **13.** > **15.** < **17.** always **19.** always

Pages 95–96   Lesson 1-10   Extra Practice

21. irrational   **23.** natural, whole, integer, rational
25. irrational   **27.** >   **29.** 18.52 m   **31.** <   **33.** >   **35.** real, rational, whole, integer, natural   **37.** 5 or −5   **39.** 0.8 or −0.8
41. 5.736×10^5

Page 99   Chapter Review   Vocabulary Check

Across
5. perfect cube   **9.** rational number
Down
1. radical sign   **3.** exponent   **7.** cube root

Page 100   Chapter Review   Key Concept Check

1. real   **3.** Power of a Product rule   **5.** 9^{28} **7.** less than

Chapter 2 Equations in One Variable

Page 110   Chapter 2   Are You Ready?

1. −17   **3.** 19   **5.** 7   **7.** $18 + h = 92$; 74 marbles

Pages 115–116   Lesson 2-1   Independent Practice

1. 72   **3** 24   **5.** −12   **7.** 2   **9.** $\frac{4}{5}$
11 q = total questions; $0.8q = 16$; 20 questions
13. Multiplicative Inverse: $1\frac{1}{3}$, $-\frac{1}{2}$; Division: 0.2, −5   **15.** true; Sample answer: The product of $\frac{3}{4}$ and $\frac{4}{3}$ is $\frac{12}{12}$, which simplifies to 1.   **17.** 53; Since $10 = \frac{1}{5}x$, then $x = 50$ and $x + 3 = 53$.

Pages 117–118   Lesson 2-1   Extra Practice

19. $1\frac{1}{4}$ **21.** $3\frac{1}{2}$ **23.** −10.5   **25.** $6\frac{3}{10}$ **27.** $\frac{8}{9}$
29.
$$-\frac{7}{8}x = 24$$
$$\left(-\frac{8}{7}\right)\left(-\frac{7}{8}x\right) = 24\left(-\frac{8}{7}\right)$$
$$x = -27\frac{3}{7}$$

31. 14.8 mi   **33.** −15   **35.** 4.5   **37.** $-7\frac{1}{2}$

Pages 125–126   Lesson 2-2   Independent Practice

1. 3   **3.** −4   **5** −52   **7** 5 bracelets   **9.** 64   **11.** −26
13a. 146 messages   **13b.** 135 messages   **15.** Sample answer: Andrea saved x dollars each week for 3 weeks. She spent $25 and had $125 left. How much did she save each week?; $50

Pages 127–128   Lesson 2-2   Extra Practice

17. 6   **19.** −3   **21.** −8   **23.** 7   **25.** −10   **27.** $1.50; Sample answer: Subtraction Property of Equality, Division Property of Equality   **29a.** no   **29b.** yes   **29c.** yes   **31.** 19   **33.** $s + 9 = 21$; 12 years   **35.** $18x = 72$; 4

Pages 133–134   Lesson 2-3   Independent Practice

1. $5n − 4 = 11$ **3** $7n − 6 = −20$ **5** s = the number of songs; $0.25s + 9.99 = 113.74$; 415 songs   **7.** s = height of the Statue of Liberty; $s + (s + 0.89) = 92.99$; 46.05 m   **9a.** $175 = 3c − 20$; 65 mph   **9b.** $s = \frac{1}{5} \cdot 175 − 1$; 34 mph   **9c.** $175 = 6h + 13$; 27 mph   **11.** $n + 2n + (n + 3) = 27$; 6, 9, 12

Pages 135–136   Lesson 2-3   Extra Practice

15. $4n + 16 = −2$ **17.** $6 + 9n = 456$ **19.** x = the number of groups of pitches; $4 + 0.75x = 7$; 4 groups   **21.** $6w + 6$; $6w + 6 = 36$; 5 units   **23.** 154   **25.** −56   **27.** $p − 14 = 17$; 31 points

Page 139   Problem-Solving Investigation   Work Backward

Case 3. $92   **Case 5.** $1,163.50

Pages 149–150   Lesson 2-4   Independent Practice

1. −2   **3** 10   **5.** 48   **7** Let n = the number; $0.5n − 9 = 4n + 5$; −4   **9a.** Sample answer: Set the side lengths equal to each other and solve for x.   **9b.** $4x − 2 = 2x + 8$ **9c.** 18 units
11. Sample answer: You have 20 crafts made and continue to make crafts at the rate of 3 per hour. How many hours will it take you and your friend to make the same amount of crafts, if she makes crafts at a rate of 5 per hour.   **13.** Sample answer: $3x + 6 = x + 7$, $1 − n = 3n − 1$

Pages 151–152   Lesson 2-4   Extra Practice

15. 3   **17.** 5   **19.** Let n = the number; $3n − 18 = 2n$; 18
21. $60x = 8x + 26$; 0.5   **23a.** True   **23b.** False   **23c.** True
23d. True   **25.** $−8y + 8$ **27.** $2z + \frac{10}{3}$

Pages 157–158   Lesson 2-5   Independent Practice

1. −9   **3** 6   **5.** null set or no solution   **7.** −6   **9** $4.72
11a. $20 + 0.15m = 30 + 0.1m$ **11b.** $m = 200$; 200 messages
13. $2(3x + 4) + 2(4x + 3) = 8(2x + 1)$; 13 in. and 15 in.

Pages 159–160   Lesson 2-5   Extra Practice

15. 13   **17.** $4\frac{4}{7}$ **19.** −9   **21.** identity or all numbers
23a. Sample answer: $3x + 5 = 3x − 2 + 7$ **23b.** Sample answer: $2(x − 1) = 2x + 2$ **25a.** False   **25b.** True
25c. False
27. $x \geq −3$

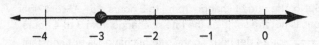

29. $n < 8$

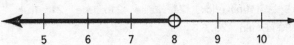

1. coefficient   **3.** properties

1. 31   **3.** Sample steps: 2; 3; 1; 4; $7\frac{2}{7}$

Chapter 3 Equations in Two Variables

1. 9   **3.** −7   **5.** 6   **7.** $\frac{2}{5}$ **9.** $\frac{1}{5}$ **11.** $-\frac{2}{5}$

1 Yes; the rate of change between cost and time for each hour is a constant 3¢ per hour.   **3.** Yes; the rate of change between vinegar and oil for each cup of oil is a constant $\frac{3}{8}$ cup vinegar per cup of oil.   **5** Yes; the rate of change between the actual distance and the map distance for each inch on the map is a constant 7.5 mi/in.   **7.** Yes; the ratio of the cost to time is a constant 3¢ per hour, so the relationship is proportional.   **9.** Yes; the ratio of actual distance to map distance is a constant $\frac{15}{2}$ miles per inch, so the relationship is proportional.
11. Sample answer:

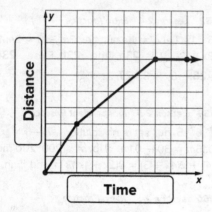

13a. no; Sample answer: $\frac{3.50}{1} \neq \frac{4.00}{2}$
13b. yes; Sample answer: $\frac{2.50}{1} = \frac{5.00}{2} = \frac{7.50}{3} = \frac{10.00}{4}$

15. No; the rate of change from 1 to 2 hours, $\frac{24 - 12}{2 - 1}$ or 12 per hour, is not the same as the rate of change from 3 to 4 hours, $\frac{60 - 36}{4 - 3}$ or 24 per hour, so the rate of change is not constant.
17. −50 mph; the distance decreased by 50 miles every hour.
19. 0.5; $\frac{1}{2}$ of retail price.   **21a.** False   **21b.** True   **21c.** True
23. 750 kB/min   **25.** 6.6 m/s

1 $-\frac{5}{8}$ **3.** $-\frac{3}{4}$ **5.** 2   **7** −4   **9.** yes; $\frac{1}{15} < \frac{1}{12}$
11. Jacob did not use the *x*-coordinates in the same order as the *y*-coordinates.
$$m = \frac{3 - 2}{4 - 0}$$
$$m = \frac{1}{4}$$
13a. Sample answer: (1, 1), (2, 6), (3, 11)   **13b.** Sample answer: (1, 1), (6, 2), (11, 3)   **13c.** Sample answer: (1, 1), (0, 6), (−1, 11)

15. 3   **17.** −3   **19.** 2   **21.** $\frac{1}{5}$ **23.** (0, 4); (4, 1); $-\frac{3}{4}$
25. $\frac{30}{180} = \frac{x}{240}$; 40 minutes   **27.** 15   **29.** 7.5   **31.** −60

1. $0.50 per paper   **3** Computers R Us; Sample answer: The unit cost for Computer Access is $25 per hour. The unit cost for Computers R Us is $23.50. 23.5 < 25   **5.** yes; 4   **7** 127 cm
9. $y = \frac{2}{5}x$; 4   **11.** Sample answer: (4, 3), (8, 6), (0, 0)
13. Sample answer: The total cost *y* of buying *x* boxes of popcorn is a proportional linear relationship. If you buy *x* boxes of popcorn and a drink for $1, the relationship between the total cost and the boxes of popcorn becomes nonproportional.

15. $y = 4.2x$; $4.20 per pound

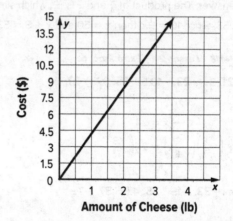

17. tickets to the play; Sample answer: The unit rate per raffle ticket is $5 and the unit rate per play ticket is $6.25. 6.25 > 5
19a. True   **19b.** False   **19c.** False   **21.** $-\frac{3}{4}$ **23.** 0   **25.** $-\frac{1}{2}$

1. 3; 4   **3.** −3; −4   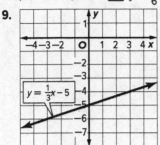 $y = \frac{5}{6}x + 8$ **7.** $y = \frac{5}{4}x - 12$

9.

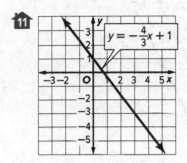

11.

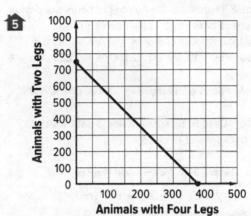

13. 0; Sample answer: A line that has a y-intercept but no x-intercept is a horizontal line.   **15.** Quadrants I, II, and IV; if a y-intercept is graphed at $(0, b)$, where b is positive, and then a line is drawn through the point so that it has a negative slope, the line will pass through Quadrants I, II, and IV.

17. −5; 2   **19.** 2; 8   **21.** $y = -2x + 3$

23.

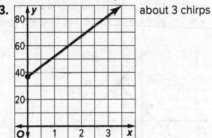   about 3 chirps

25. $y = 8x + 5$ **27.** 8; 20; y-intercept; slope   **29.** −2
31. yes; $\frac{4}{1}$ or 4

1. x-intercept: 3.5; y-intercept: 7

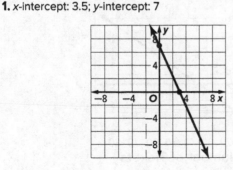

3 x-intercept: $1\frac{1}{4}$; y-intercept: $1\frac{2}{3}$

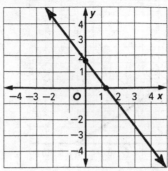

5

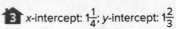

The x-intercept of 375 means that if the zoo had only four-legged animals, there would be 375 of them. The y-intercept of 750 means that if the zoo had only two-legged animals, there would be 750 of them.   **7.** After $3x = 12$, Carmen didn't divide both sides by 3 to get the x-intercept of 4.   **9.** Sample answer: $x = 2$; $y = 2$

11. (12, 0), (0, 8)

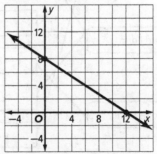

13. (6, 0), (0, 10)

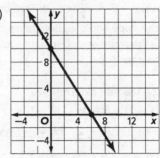

15.

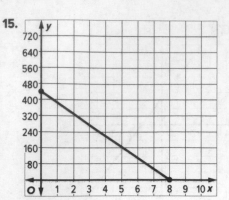

The x-intercept, 8, represents the number of hours the painter worked to finish the basement. The y-intercept, 440, represents the total amount of money she has to pay the painter. **17.** $50; -60$ **19.** $2x + 1$ **21.** $4x + 6$ **23.** $-6a - 6$

Page 219 Problem-Solving Investigation Guess, Check, and Revise

Case 3. 53 rolls **Case 5.** Sample answer: 3 packages of 8 cards and 4 packages of 12 cards

Pages 225–226 Lesson 3-6 Independent Practice

1. $y - 9 = 2(x - 1); y = 2x + 7$ **3.** $y + 5 = \frac{3}{4}(x + 4);$ $y = \frac{3}{4}x - 2$ **5** Sample answer: $y + 4 = -\frac{3}{2}(x - 4);$ $y = -\frac{3}{2}x + 2$ **7.** Sample answer: $y - 14 = \frac{1}{5}(x - 10)$ **9** $3x + y = 13$

11.

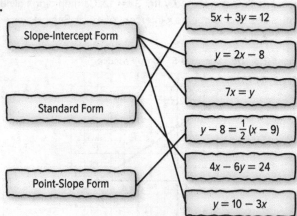

Slope-Intercept Form

Standard Form

Point-Slope Form

$5x + 3y = 12$

$y = 2x - 8$

$7x = y$

$y - 8 = \frac{1}{2}(x - 9)$

$4x - 6y = 24$

$y = 10 - 3x$

13. Sample answer: $y - 5 = -\frac{1}{2}(x - 2)$; First, use the equation to find the slope and the coordinates of any point on the line. Then use the slope and coordinates to write an equation in point-slope form.

Pages 227–228 Lesson 3-6 Extra Practice

15. $y - 10 = -4(x + 7); y = -4x - 18$
17. $y - 2 = \frac{2}{3}(x - 6); y = \frac{2}{3}x - 2$ **19.** $4x - 5y = 17$
21. Sample answer: $y - 3 = -\frac{5}{2}(x + 2)$ **23.** $y = 4x - 2;$ $y - 2 = 4(x - 1)$ **25.** \$9 per hour **27.** $d = 120t$; 600 mi

Pages 239–240 Lesson 3-7 Independent Practice

1. $(4, 4)$

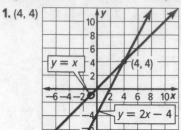

3 no solution

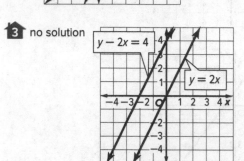

5. $(0, 3)$

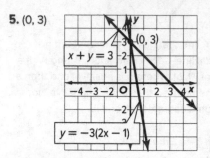

7. Sample answer: Let $x = $ the number of dogs and $y = $ the number of cats; $x + y = 45, y = x + 7$; There are 19 dogs and 26 cats.

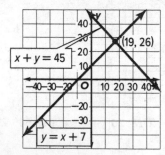

9. one solution

11 a.

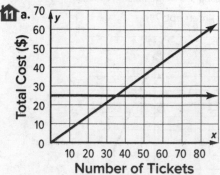

11b. 18 rides **13.** sometimes; Sample answer: $y = 2x + 1$ and $y = 5x + 1$ intersect at $(0, 1)$ and $b = d$. However, $y = 2x + 1$ and $y = x + 2$ intersect at $(1, 3)$, but $b \neq d$.

Pages 241–242   Lesson 3-7   Extra Practice

15. (1, 2)

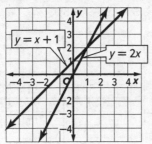

17. no solution

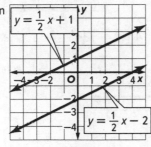

19. (−2, 0); one solution   **21a.** one solution   **21b.** no solution   **21c.** one solution   **23.** $y = 5$ **25.** $x = -28$

27. $y = \frac{1}{4}$

Pages 247–248   Lesson 3-8   Independent Practice

1. (1, 6)   **3** (−2, −12)   **5.** (7, 11)   **7.** $\left(\frac{1}{2}, 12\frac{1}{2}\right)$ or (0.5, 12.5)

9. Sample answer: $s + p = 15$; $p + 7 = s$; (4, 11); She bought 11 shirts and 4 pairs of pants.   **11** Sample answer: $8x + 2y = 18$; $3x + y = 7.50$; (1.5, 3); A muffin costs $1.50 and 1 quart of milk costs $3.   **13.** ∅; Sample answer: Adding $5x$ to each side of $y = -5x + 8$ results in the equation $5x + y = 8$. Since $5x + y$ cannot equal both 8 and 2, there are no values for x and y that make this system of equations true.   **15.** $y = x + 3$ and $y = -2x - 3$; Sample answer: The solution of $y = x + 3$ and $y = -2x - 3$ is (−2, 1). The solution of the other three systems is (1, −2).

Pages 249–250   Lesson 3-8   Extra Practice

17. (−12, −3)   **19.** (−3, 0)   **21.** (5, −1)   **23a.** $y = 4x + 95$ and $y = 9x + 75$

23b.

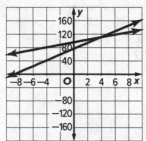

(4, 111); the costs, $111, are the same if 4 students attend either.   **23c.** nature center   **25.** $n = 3v$, $n + v = 20$; Naomi: 15 spikes, Vicki: 5 spikes   **27.** 67   **29.** 22.9   **31.** 97   **33.** 78

Page 255   Chapter Review   Vocabulary Check

Across

5. direct variation   **7.** x intercept

Down

1. substitution   **3.** y intercept

Page 256   Chapter Review   Key Concept Check

1. $y = -0.5x + 1$ **3.** $y = 0.5x$ **5.** $x = 5$

Chapter 4 Functions

Page 266   Chapter 4   Are You Ready?

1. (1.5, 2.5)   **3.** (0, 1.5)   **5.** (1, 1)   **7.** −18   **9.** −3   **11.** $901

Pages 273–274   Lesson 4-1   Independent Practice

1 **a.** $b = 45d$; Forty-five baskets are produced every day.

b. 16,425 baskets   **3** **a.** $f = 3.5 + 0.15d$

b.

d	$3.5 + 0.15d$	f
10	3.5 + 0.15(10)	5.00
15	3.5 + 0.15(15)	5.75
20	3.5 + 0.15(20)	6.50
25	3.5 + 0.15(25)	7.25

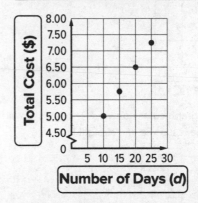

5. Sample answer: $d = 60t$; A car is traveling at a rate of 60 miles per hour.

Pages 275–276   Lesson 4-1   Extra Practice

9a. $d = 15w + 5$ **9b.** $365   **11.** 32 text message cost $4.80; 50 text messages cost $7.50; 70 text messages cost $10.50   **13a.** $y = 28.4x$ **13b.** 4,260 g

Pages 281–282 Lesson 4-2 Independent Practice

1 D: {−6, 0, 2, 8}; R: {−9, −8, 5}

x	y
8	5
−6	−9
2	5
0	−8

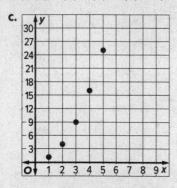

3.

x	825x	y
1	825(1)	825
2	825(2)	1,650
3	825(3)	2,475
4	825(4)	3,300
5	825(5)	4,125

5 a. To get the y-value, the x-value was multiplied by itself.
b. (1, 1), (2, 4), (3, 9), (4, 16), (5, 25)

c.

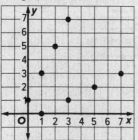

d. Sample answer: This graph curves upward. The points in all of the other graphs in the lesson lie in a straight line.

7a, 7c.

b. (1, 0), (3, 1), (5, 2), (7, 3) **d.** Sample answer: The distance between each point in the original table and the x-axis is the same as the distance betweeen the points with the reversed ordered pairs and the y-axis. **9.** Sample answer: The domain is the set of x-coordinates. Morgan listed the set of y-coordinates; {−4, 0, 1, 2}

Pages 283–284 Lesson 4-2 Extra Practice

11. D: {−1.5, 2.5, 3}; R: {−3.5, −1.5, −1, 3.5}

x	y
−1.5	3.5
2.5	−1.5
3	−1
−1.5	−3.5

13. 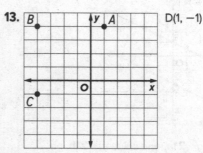 D(1, −1)

15a. True **15b.** True **15c.** True **17.** $\left(\frac{3}{4}, \frac{1}{2}\right)$ **19.** $\left(1, -\frac{3}{4}\right)$
21. $\left(-\frac{1}{2}, -\frac{1}{2}\right)$ **23.** $\left(-1, \frac{1}{4}\right)$

Pages 291–292 Lesson 4-3 Independent Practice

1. 35 **3** 11
5 Sample answer:

x	5 − 2x	f(x)
−2	5 − 2(−2)	9
0	5 − 2(0)	5
3	5 − 2(3)	−1
5	5 − 2(5)	−5

D: {−2, 0, 3, 5}
R: {9, 5, −1, −5}

7a. The total points p(g) is the dependent variable and the number of games g is the independent variable. **7b.** Only whole numbers between and including 0 and 82 make sense for the domain because you do not want data for a partial game and there are only 82 games in a season. The range will be multiples of 20.7. **7c.** p(g) = 20.7g; 186.3 points **9.** 2 **11.** Sample answer: f(x) = 2x − 2; f(0) = −2, f(−4) = −10, f(3) = 4
13a. 101 **13b.** 138 **13c.** −982

Pages 293–294 Lesson 4-3 Extra Practice

15. −41
17. Sample answer:

x	x − 9	f(x)
−2	−2 − 9	−11
−1	−1 − 9	−10
7	7 − 9	−2
12	12 − 9	3

D: {−2, −1, 7, 12}
R: {−11, −10, −2, 3}

19b.

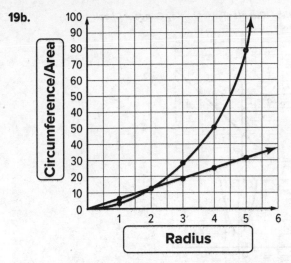

19c. Circumference: linear; sample answer: When the ordered pairs are graphed, the points fall in a line. Area: nonlinear; sample answer: When the ordered pairs are graphed, the points do not fall in a line. **21a.** False **21b.** False
21c. True **23.** −14 **25a.** $c = 5d$; Riley makes an average of 5 phone calls per day. **25b.** 35 phone calls

Pages 339–340 Lesson 4-8 Independent Practice

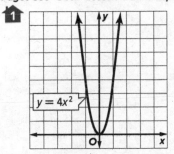

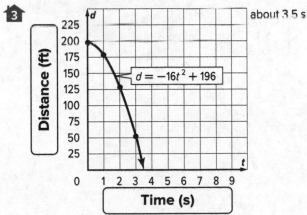

about 3.5 s

5a. $A = 12x - x^2$ **5b.** 6 in. by 6 in. **7.** nonlinear; Sample answer: The function is quadratic. **9.** linear; Sample answer: The equation is written in slope-intercept form so it is a straight line. **11.** nonlinear; Sample answer: The function is quadratic. **13.** Sample answer: $y = x^2 - 3.5$

Pages 341–342 Lesson 4-8 Extra Practice

15. $y = -x^2 + 2$

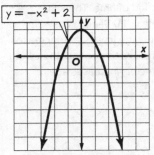

17.

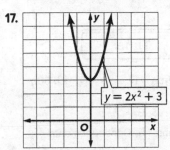

19a. $A = x^2 + 4x$
19b.

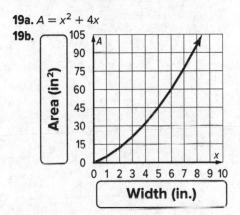

19c. 96 in² **21a.** True **21b.** False **21c.** True **23.** 1,296
25. 243 **27.** 16 **29.** 36 **31.** 1 **33.** 3

Pages 351–352 Lesson 4-9 Independent Practice

1 Sample answer: Luis starts out from his home. He walks away from his home, stops to let the dog run around, and walks further away from home. Then he walks towards home.
3 Sample answer:

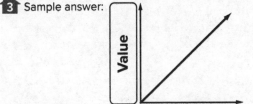

5a. Sample answer: Hector hikes at a steady rate. **5b.** Sample answer: Hector suddenly stops hiking. **5c.** increase; Sample answer: The graph rises from left to right at the beginning.
7. Graph A; Sample answer: Graph A increases from left to right at a constant rate then levels off. This represents a tree growing steadily before it stops growing.

9. Justine rode her bike at a constant rate in the beginning. She then stopped riding for a period of time. Then she continued riding at a constant rate.   **11.** Sample answer: Mrs. Fraser's electric bill starts out high in January, increases until about March, and then decreases throughout the spring and summer. In the fall, the electric bill increases again.

13. Sample answer:

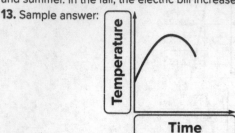

15a. True   **15b.** False   **15c.** True   **17.** $2p + 20$ **19.** $60q$

1. relation   **3.** qualitative graphs   **5.** Continuous data
7. function

1.

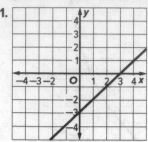

3.

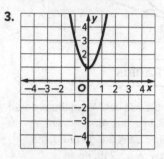

Index

Index

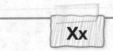

$$=$$

Equation Mat **WM1**

Work Mats

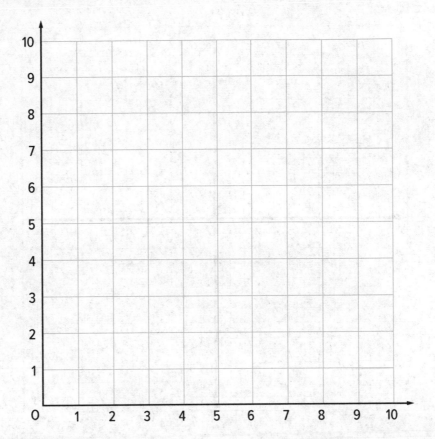

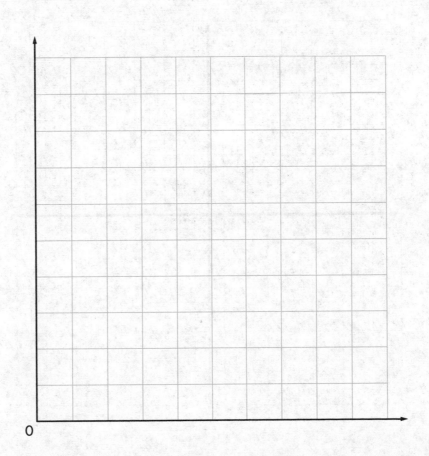

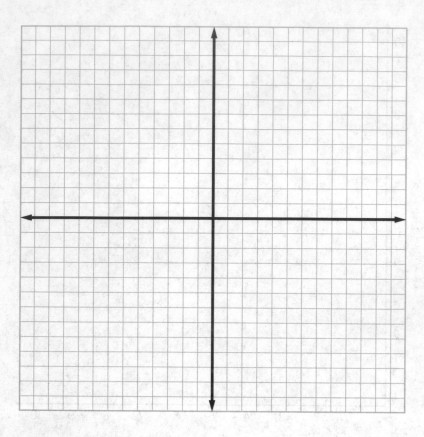

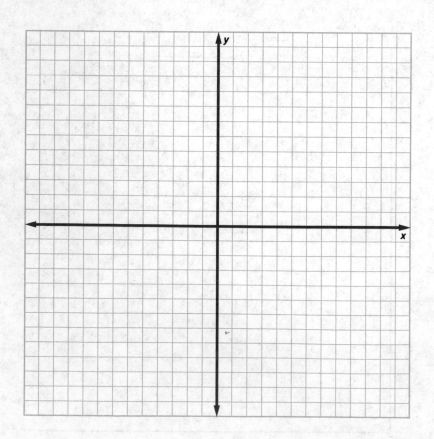

What Are Foldables and How Do I Create Them?

Foldables are three-dimensional graphic organizers that help you create study guides for each chapter in your book.

Step 1 Go to the back of your book to find the Foldable for the chapter you are currently studying. Follow the cutting and assembly instructions at the top of the page.

Step 2 Go to the Key Concept Check at the end of the chapter you are currently studying. Match up the tabs and attach your Foldable to this page. Dotted tabs show where to place your Foldable. Striped tabs indicate where to tape the Foldable.

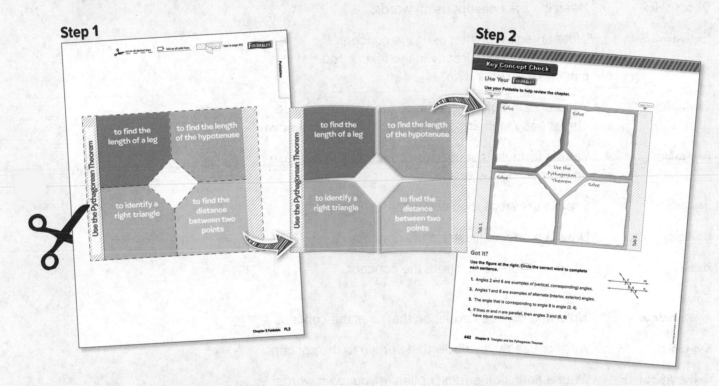

How Will I Know When to Use My Foldable?

When it's time to work on your Foldable, you will see a Foldables logo at the bottom of the **Rate Yourself!** box on the Guided Practice pages. This lets you know that it is time to update it with concepts from that lesson. Once you've completed your Foldable, use it to study for the chapter test.

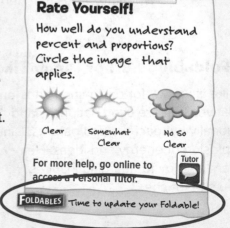

How Do I Complete My Foldable?

No two Foldables in your book will look alike. However, some will ask you to fill in similar information. Below are some of the instructions you'll see as you complete your Foldable. **HAVE FUN** learning math using Foldables!

Instructions and what they mean

Best Used to...	Complete the sentence explaining when the concept should be used.
Definition	Write a definition in your own words.
Description	Describe the concept using words.
Equation	Write an equation that uses the concept. You may use one already in the text or you can make up your own.
Example	Write an example about the concept. You may use one already in the text or you can make up your own.
Formulas	Write a formula that uses the concept. You may use one already in the text.
How do I ...?	Explain the steps involved in the concept.
Models	Draw a model to illustrate the concept.
Picture	Draw a picture to illustrate the concept.
Solve Algebraically	Write and solve an equation that uses the concept.
Symbols	Write or use the symbols that pertain to the concept.
Write About It	Write a definition or description in your own words.
Words	Write the words that pertain to the concept.

Meet Foldables Author Dinah Zike

Dinah Zike is known for designing hands-on manipulatives that are used nationally and internationally by teachers and parents. Dinah is an explosion of energy and ideas. Her excitement and joy for learning inspires everyone she touches.

Laws of Exponents

Product of Powers

Quotient of Powers

Power of Powers

cut on all dashed lines

fold on all solid lines

tape to page 100

FOLDABLES

Examples

Examples

Examples

page 100

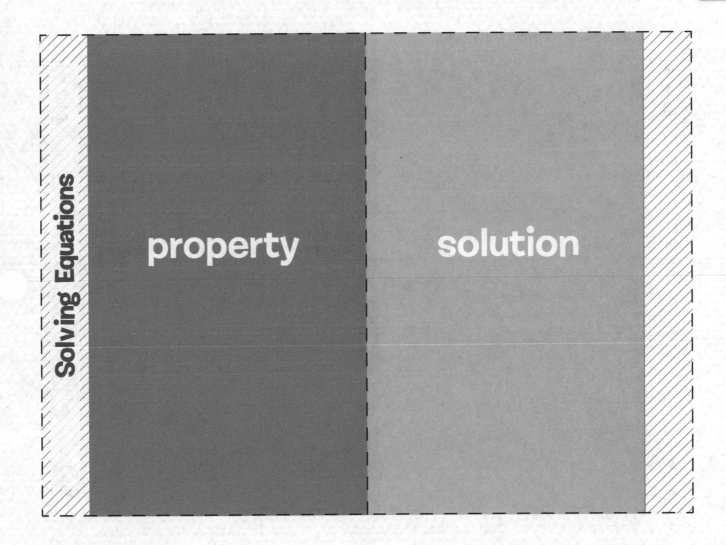

Solving Equations

property

solution

Foldables

cut on all dashed lines | fold on all solid lines | tape to page 164

FOLDABLES

Tab 2 (page 164)		Tab 1 (page 164)
	Step 1	Distributive Property
	Step 2	Addition or Subtraction Property of Equality
	Step 3	
	Step 4	Multiplication or Division Property of Equality

Solve Systems of Equations

one solution	no solution	infinite number of solutions

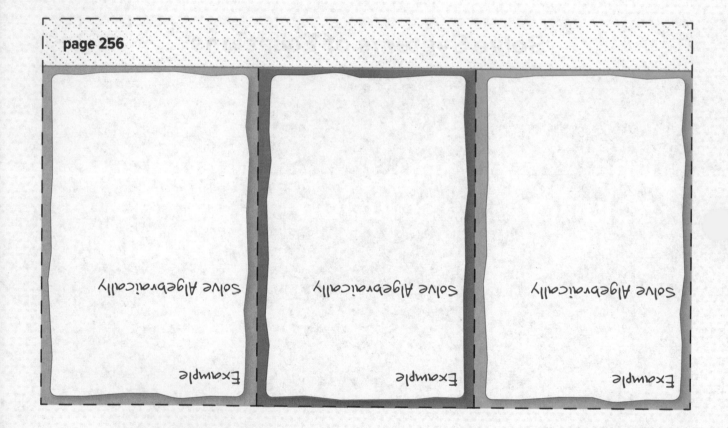

page 256

Solve Algebraically

Example

Solve Algebraically

Example

Solve Algebraically

Example

Relations and Functions

relations

functions

linear

nonlinear

cut on all dashed lines   fold on all solid lines   tape to page 358   FOLDABLES

page 358   Tab 3

Examples

Words

page 358   Tab 2

Examples

Words

page 358   Tab 1

Example

Example

Words

Words

FL10   **Chapter 4 Foldable**